Writing and Publishing Scientific Articles: A Complete Guide

Writing and Publishing Scientific Articles: A Complete Guide

Gweneth Glover

Larsen & Keller
www.larsen-keller.com

Writing and Publishing Scientific Articles: A Complete Guide
Gweneth Glover
ISBN: 979-8-88836-039-2 (Hardback)

Larsen & Keller
Published by Larsen and Keller Education,
5 Penn Plaza,
19th Floor,
New York, NY 10001, USA

Cataloging-in-Publication Data

Writing and publishing scientific articles : a complete guide / Gweneth Glover.
p. cm.
Includes bibliographical references and index.
ISBN 979-8-88836-039-2
1. Academic writing. 2. Technical writing. 3. Scholarly publishing. 4. Academic writing--Guidebooks. 5. Technical writing--Guidebooks. 6. Scholarly publishing--Guidebooks. I. Glover, Gweneth.
P301.5.A27 W75 2023
808.066 378--dc23

For more information regarding Larsen and Keller Education and its products, please visit the publisher's website www.larsen-keller.com

Table of Contents

Preface

Scientific article present the findings of a research and must follow a standard structure of a scientific manuscript. This structure must be followed in all aspects including the references and content of the manuscript. It offers significant contributions to the body of knowledge on topics that are interesting and pertinent to a wide range of readership groups and constituencies including research practitioners, public policymakers, the general public, managers, marketing educators and other academic researchers. Scientific articles must be written concisely in order to be more understandable. Scientific writing and publication are critical for advancing public health knowledge and practice. This book is compiled in such a manner, that it will provide in-depth knowledge about the theory and practice of writing and publishing scientific articles. It is an essential guide for both academicians and those who wish to learn the craft of writing scientific articles.

This book has been the outcome of endless efforts put in by authors and researchers on various issues and topics within the field. The book is a comprehensive collection of significant researches that are addressed in a variety of chapters. It will surely enhance the knowledge of the field among readers across the globe.

It gives us an immense pleasure to thank our researchers and authors for their efforts to submit their piece of writing before the deadlines. Finally in the end, I would like to thank my family and colleagues who have been a great source of inspiration and support.

Gweneth Glover

PART I

Before Beginning the Paper

Lectori Salutem

My reader, allow me to greet you with the words of the Latin writers: *lectori salutem*. You are holding a book that, while it cannot claim to be unique, distils many years of experience, spanning virtually my entire career as a scientist, publishing author, and editor. As a young scientist, eager to publish internationally, the book that first caught my attention in the field of scientific writing was Robert Day's *How to Write and Publish a Scientific Paper*. I have used this book widely in its various editions, and I am glad to record my gratitude to this author for his fine book (now, in the latest editions, with co-author Barbara Gastel).

As Gastel and Day (2016) correctly observe, scientific writing is a rather rigidly regulated area of writing. Consequently, any book aiming to provide advice in this area will resemble others. Why, then, is there a need to write about this again and again?

My reply to that question is that I found Day's book too closely tailored to the traditions and views of the North American scientific community and, despite the occasional nod to acknowledge alternative traditions in publishing, they did not really aim to enlighten non-native English speakers. This shortcoming still characterises the latest edition (Gastel and Day, 2016). True, there are a lot of similarities, perhaps more than there are differences. However, "non-native" scientists working and writing in a different environment have a different view and, perhaps, would benefit from the approach of a non-native writer, whose own publication record is in mostly non-U.S.-based forums.

One area where my advice deviates considerably from Gastel and Day's (2016) is on scientific figures. This is more than a slight difference of opinion—it seems a different philosophy. I confess to adhere to the principles advocated by William Cleveland and Edward Tufte and find much to lament about the current standard of figures, even in the most prominent scientific journals. This field is in dire need of more attention and the practice of designing figures would benefit richly from a more attentive approach. Thus, I place significant emphasis on constructing figures for both analysing and presenting data.

In general, though, this is not a "how-to" book. Allow me to use an analogy: You can possibly learn to swim when thrown into the water, with the trainer standing at the edge of the pool, explaining the motions to make. She will certainly have your full attention. I believe that this "learning while doing" method has some merits. However, it just may be of use if you first familiarise yourself with the swimming pool area: the general setup, the types, kinds and features of different pools, where to get into the various pools and how to get out, the water temperature and depth in each, where to go if you want to change, how to get help, and so on. Only then, of course, should you jump in. This book follows the second approach, and seeks to inform you about the publication process itself, including information on journal types, as well as the process of scientific and technical editing. I believe that knowing the whole process by which your manuscript will become a published paper can help you to navigate this process more effectively, less painfully, and — of utmost importance to scientists — *faster.*

In this volume, I discuss aspects related to writing and publishing different kinds of scientific papers. Most of the emphasis will be on the so-called primary scientific paper, whilst shorter chapters detail special features of reviews, conference proceedings papers, and book chapters. My approach is also distinct in that I do not extensively discuss the elements of style. There are many good books available on this subject (Turabian, 2007; Barrass, 2015; Cargill and O'Connor, 2013). Given my own background, most of my examples come from environmental science in the broad sense. However, with extensive teaching experience, I can claim that scientists from various other fields, from economics to social sciences, have found the material usable and useful.

Another caveat: I assume that my reader has, first of all, valuable scientific results and her interest is in *how* to present them to best effect. In other words, my reader has some publishable results which she trusts. This book is not to help the confused, who have generated a lot of data, and do not know what to do with them. Secondly, I do not offer a kind of "cookbook", with recipes detailing how to get your results published. I shall provide some guidance but there is no guarantee that, if you follow these points, your manuscript will be published in the first journal you submit to, and will be published quickly. Rather, my philosophy is different: I try to instil an attitude (see Chapter 1), so that you see the publication process more in perspective, and I urge you to pay attention to the work others are doing on your manuscript — this consideration will pay off handsomely. I hope you will find this approach helpful.

During the last 25+ years, I have gained much from teaching courses on scientific writing to students and scientist with widely different cultural backgrounds, from Hungary to Denmark, and from China to Burkina Faso. I am grateful to the participants of these courses for their enthusiasm, probing questions, and feedback. I also thank many colleagues who inspired me with their discussions, comments and papers, especially my dear friend Dr Ferenc Szentkirályi, who first suggested that I hold a course on scientific writing; to my colleagues at the Training Centre in Communication, Nairobi, Kenya and to Drs Søren Toft, Éva Vincze, Miklós Sárospataki, Marco Ferrante, Béla Tóthmérész, Judit Fazekas, László Gallé, Erzsébet Hornung, Jørgen Jakobsen, László Körmöczi, Fang-hao Wan, Min-sheng You, Nian-wan Yang, Eric Danso, Paolo A.V. Borges, Margarida Matos, and Eve Veromann, as well as Lene Gregersen, Anne Olsen, Karen Konradi, and Warwick Thomson for comments and support. Special thanks to Ms Joy Owango, with whom we established the Training Centre in Communication, devoted to such training in Africa, who then worked tirelessly to develop this endeavour in Kenya, and now in several other African countries.

When writing this book, I have had support from the former Danish Institute of Agricultural Sciences, now Aarhus University (Denmark), for which I am grateful. I thank Drs Zoltán Elek and Marco Ferrante who helped me to produce several of the figures. I also thank warmly Open Book Publishers, especially Dr Alessandra Tosi, for their trust in this book and for a multitude of editorial help, from organising reviews to designing a cover; Drs Richard Primack, Miklós Sárospataki, and John Wilson for helpful comments, Dr Clive Nuttman for linguistic review, Ms. Lucy Barnes and Melissa Purkiss for editorial assistance.

Notwithstanding the assistance, advice and encouragement from many people, this book is an individual interpretation of some common rules of scientific publishing. I tried to present my arguments supported by facts, experience and examples. Any bias, error or omission remains solely my own.

1

A Few Fundamentals

Arpád Tóth, Hungarian poet, bemoaned, in a heart-rending poem written in 1923, that there is no direct route "from soul to soul". While his pessimism is perhaps not totally warranted, he was right. When we have an idea, a discovery, and want to let others know, we have to bridge a gap between ourselves and others: we must express ourselves somehow. Trying to grab others, take them where we saw the new fact, and let them see for themselves is not usually enough nor practicable. The human race has long abandoned this as the sole, or chief, method of passing on acquired knowledge. We have invented different systems that have one thing in common: we have to use some system to *code* the information we possess. Language is one of those codes. The initiator must understand the new information then learn a coding system to express herself. However, even though this is not easy, it is only the first half of the communication process. The rest is the receiver's task: to *decode* the information in order to understand it. Thus, for effective communication, *two* processes must be conducted successfully. The initiator's responsibility is bigger here, as she can make the decoding process really, even if often involuntarily, difficult. She can also try her best to help the receiver, i.e. the person(s) trying to decode and understand the information.

It is very common that we, when in the role of initiator, use the coding system differently when we face different receivers. For example, when we talk about our scientific work, we instinctively use different words (modify the code), depending on whether we are speaking with our peers, friends, family, or our children. Why? Because we are aware of their differences in levels of expertise in the subject, in vocabulary, in attitude. In other words, we *modulate* our message, our use of coding, to try to meet the expectations and level of skill of our communication partners. We try to make the decoding easier for them. So, effective communication requires successful coding *as well as* decoding. These processes are done by different people, who are often not in personal contact, thus the two processes are typically conducted separately. The receiver is often on her own, to interpret (decode) the message successfully.

From this follows the most important principle of communication: *the simpler the code, the easier is the understanding*, and the lower are the chances of misunderstanding. So, the coding of information not only has to be precise — it also must be expressed using *the simplest possible code*. However, there is a conflict here: precise coding is often very complicated. As we move from the complicated to the simple, the message will lose precision, and will also lose complexity and articulation. We can move along the complexity-simplicity continuum towards simplicity only so far, before the message will become too simple. At such oversimplification, the message will become so general as to be meaningless. Therefore, we should aim for simplicity, but we should very judiciously guard how far we go in this process. Nonetheless, the most frequent problem in scientific communication is unnecessary complexity, rather than oversimplification.

A special point here is that we want to present new discoveries — facts that, so far, no one has established. Understanding novelty is a challenge, so we best serve the receiver if we do not unnecessarily make her task more difficult by using a very complicated code. If communicating successfully requires that we use the simplest appropriate coding, then in the case of presenting new discoveries, the use of the simplest possible coding is mandatory.

Actually, we are in luck. The fact that today's scientific *lingua franca* is English helps us. Why? Because English is a language where simplicity is also a stylistic virtue, so we can write both effectively and elegantly by using uncomplicated language.

So, all elements are consilient: we have data that we genuinely trust are new, exciting and reliable. We would like to show them to the wide world. When presenting new discoveries, simplicity eases understanding; it also increases elegance.

A final note: scientific publishing is not only about ourselves. It is not merely to show off, as it were, our ingenuity to the world, our discovery of a number of "firsts". It is to start a dialogue, to invite feedback. It is a continuing dialogue about the world, about its facts, rules, and exceptions. About its beauty, and about the beauty of the call of the unknown. A quest that has no end, and whose reward is not an occasional rest on some peak, or plateau that we have reached, however small. Rather, the reward is the unending vista, the continuing pursuit. It is the voyage with no arrival.

2

Types of Publications and Scientific Literature

Why Do We Publish?

If a tree falls in a forest, and no one is there to hear it, does it make a sound?

Day (1998) suggests this question as a starting point for students to understand the importance of publication in science. It is a good starting point, and most students' first response is: "Yes, it does". However, this immediate answer happens to be wrong. Few people notice first (although realise on reflection) that, by definition, sound is not equal to "pressure waves in the air". When a tree falls, it generates air pressure waves. This is not a sound. This is the physical phenomenon that is a necessary condition for a sound — but the other half, the receiver, is missing. Sound is a sensory experience that is generated in an organism's receptor organ by the air pressure wave. The pressure waves in the air become sound when someone is present who does sense (hear) it. A "sound" assumes a perception of those waves by an organism.

Following a similar line of reasoning, we can ask: if someone discovers a law of nature, proves it in well-designed experiments, the results of which are carefully documented, and then puts this document into her drawer, never to publish or show anyone, does this information exist? The law of nature in question existed before the experiments were carried out, and continues to exist independently (provided you believe in objective reality), but it only becomes "information" when we know about it, when we become *informed* about it, typically through the publication of a peer-reviewed scientific manuscript. Knowledge (or information) that is not published does not, effectively, exist, because there is no one who knows about it, and can use it. If it never comes to light (i.e. the information is not published), then doing the experiments, and (however carefully) documenting the results, was pointless and of little use — the law must be rediscovered by others. It only becomes existing information if there is a realistic chance that others can know about it. Publication is therefore a necessary, integral part of scientific research, not something that comes after the experiments (the "science") were finished. The more one can integrate the experiments with their description for publication, the easier the task, and the more convincing the results will be.

When something is published, we refer colloquially to it as "a known fact"; however, if we reflect, we realise that there is no piece of information that is universally known. Even putting aside the practical difficulties of access, knowledge of language, etc., what about people who are illiterate? What about infants, the intellectually disabled, those who, perhaps, due to an accident, suffered brain damage, or the dead or the unborn? Do we exclude them as members of humanity? No, we do not. Thinking along these lines leads us to realise that the conditions according to which we generally consider something "known" have always been a matter of agreement. If a publication (in the form required by that tradition) fulfils the criteria posed by the agreement, we generally consider the discovery as having been "validly published", and, thus, existing knowledge, irrespective of how many people actually know it, or understand it. Einstein's theory of relativity is probably not understood by most of us, yet there is no doubt about attributing the concept to his name. Today, most scientific discoveries are published in scientific journals. The chief criterion for such publication is that experts evaluate the reliability and novelty of the findings described. This "peer review" is a necessary step to valid publication of a new discovery.

This tradition of publishing new scientific discoveries/knowledge grew out of the practice of scientific societies, established at the time of the Enlightenment, when scientific research in Europe started to blossom. Members of those societies met regularly to discuss their discoveries. The first scientific journals started as the written official records of the meetings of such societies. The Proceedings of the Royal Society of London, one of the oldest scientific journals, is one such journal, and it occasionally still publishes a brief record of a discussion, when the article is based on an oral presentation. This is a remnant of the old practice when an oral presentation was followed by a discussion, during which fellow scientists present at the talk discussed the reliability of the discovery presented. Once the consensus emerged that the reported facts could be trusted, the presenter was credited with the discovery, and this was, from now on, considered "existing knowledge". Since then, scientific publishing has changed considerably, and many different types of scientific publications have emerged.

Types of Publications

Publications take many different forms, but types of scientific publication fall into fewer categories. I do not attempt an overall, definitive taxonomy of scientific communication types, and only restrict this to a few important distinctions.

The different types of theses, connected to higher education, are ones that most scientists must get acquainted with. This is often the first type of publication they themselves will have to write. Another common written publication type is the scientific article, most commonly meaning the "proper" (primary) article. In addition, there is the short communication, review article, invited article, note, comment, letter, and so on. Journals often publish book and software reviews. Scientists also write papers for conference proceedings, reports for different organisations, book chapters and books. I will briefly survey these later, but the main intention of this book is to help the reader to write a so-called primary scientific paper (hereafter "paper").

Primary scientific papers are published by specialist journals. The three general types of these are: the "society" journals, the commercial scientific journals, and the small, specialist journals. These are similar in that all of them publish peer-reviewed, primary scientific papers. This similarity of their shared primary purpose hides significant differences.

Society Journals

Most fields of science have an international society. These societies are typically not-for-profit entities, and their aim is to promote their respective fields of science. One of the ways they do this is by publishing an international scientific journal. These journals are, as a rule, of high scientific quality, well-produced, inexpensive (especially in comparison to the 'commercial' journals) and, consequently, accessible and widely circulated. The key to a high level of science published in the papers is the good peer review process that is provided by the wide range of volunteer experts in the field, who are members of the society. Members of the society can subscribe to the journal at very favourable rates and, because these societies have thousands of members, this provides the publisher with a wide circulation, ensuring that the journal is widely available in the academic community, and many people read it. Due to this wide circulation (and high scientific prestige), many scientists will send manuscripts for publication, so the journal can afford to publish only the very best of them. As the circulation is high, the subscription prices (for members of the relevant society) can be kept relatively low, because, even at low profit margins, the journal will generate significant profit. The societies and the publisher share these profits. (Societies may periodically re-negotiate the publishing contract and may change publisher).

These journals are usually the most coveted ones of the field to publish in, and they usually sit near the top of the citation statistics summaries. Examples of such journals include (in parentheses, the parent society): *Ecology* (Ecological Society of America), *Journal of Animal Ecology* (British Ecological Society), *American Naturalist* (American Society for Naturalists), *Journal of Experimental Biology* (Society of Experimental Biology), *New Phytologist*, and *Weed Research* (European Weed Research Society).

Commercial Journals

The commercial success of the top scientific society journals lured several commercial publishers into the field. Several publishers (e.g. Elsevier, Springer) have aggressively pursued the policy of founding new journals. Some of these journals took root and became respected in their fields, whilst others merged, or disappeared. The principles for editing a successful scientific journal are the same as in the case of the society-run journals. The main difference, however, is that these journals are not published as a non-profit venture to promote a field of science. The peer review is usually organised by using a smaller number of experts, often dependent on the personal networks of the editorial board members and, thus, it is not always possible to find top-level experts for every manuscript. Further, these journals are published for profit. Consequently, they are very expensive, which often results in limited circulation. A good example of commercial scientific journals is the group of journals published by Springer Verlag (including *Oecologia, Planta, Molecular and General Genetics, Theoretical and Applied Genetics,* among many others). These journals, more often than not, occupy the middle range in citation statistics.

Small Specialist Journals

There are numerous scientific journals published by small, specialised institutes, such as museums, research institutes, university departments, and botanical or zoological gardens. Such journals are of high quality within their specialist fields, but there are several unfavourable aspects of such publications. Their narrow specialisation means that the potential readership is small. Consequently, the circulation cannot be as high as for the more general journals and, thus, such journals must be relatively expensive. Moreover, the institute often has to fund the publication itself and, at times, this is not its highest priority. Consequently, the publication may appear infrequently, perhaps even irregularly. Thus, although the peer review in the journal is good, often very good, such journals, due to mostly logistical and economic factors, are of limited general impact. Examples include *Proceedings of the Missouri Botanical Garden, Annales Musei Nationalis Hungariae* (published by the Hungarian Natural History Museum), etc.

The boundaries between these categories have started to blur. Today, many societies and small institutes entrust the production of their journals to professional publishing houses. The publishers are also "on the hunt" for journals that they can buy and take over, or secure a contract to publish on behalf of a society, because it is much easier to continue, or re-organise, an existing journal than to start a new one. The main differences still remain: that the best and widest reviewer range is available to the editors of the society journals and, thus, they can have the best quality specialist reviewers. High-quality peer review is the key to a journal's quality. The society's members also provide a secure subscriber base and, hence, the society can negotiate a favourable subscription rate for its journal with the publishing house. The publishing house benefits from the large pre-existing subscription base.

Internet Publishing and Internet-Based Publications

The Internet has had a significant impact on scholarly publishing. The electronic processing of manuscripts has become near-universal. Several major publishing houses operate "electronic manuscript handling centres" that are intended to speed up the publication process. While the principles of electronic scholarly publishing are not different from the paper-based publications, there will be further changes that need to be followed by scientists aspiring to publish.

Many journals today are published electronically as well as in printed form, but there is an increasing number of journals that are only published electronically — for example, the Public Library of Science (PLoS) range of journals. From a commercial point of view, the enthusiasm of publishers to embrace "free" electronic journals is understandable. These journals are free to readers — but not free to the authors wishing to publish in these journals. As electronic publishing still costs money, the publishers, if they publish an Open Access journal, can have a faster return on their investment, because the authors pay these costs before their article is published. Income security is also improved: the profit no longer depends on the number of readers or subscribers — while in traditional publishing, it does. That said, there are other publishing models in existence, and some of them are "truly free"; it may be worth checking our the list on www.doaj.org (read attentively, because not all journals on that list are free for aspiring authors).

When considering a submission to an electronic -only journal, it is wise to think about similar factors as for the "traditional" journals (see above). Do not publish in a journal that is not well-known and respected in its field. No matter how quick, or cheap, the publication is, publishing in an electronic journal of no reputation is a waste of your work (and money). If you are careless, your work will not even be published. Several unscrupulous "publishers" will happily collect your money and never publish your paper. Avoid this disappointment.

On Predatory Journals

Scientific publishing is a profitable activity, and the possibility of making large profits has lured more than a few unscrupulous players to the field. They range from publishers who run journals with "very soft"" (uncritical) peer review, to those that have little or no quality control, those that publish all submitted manuscripts unedited and also those that trick prospective authors to part with their money but never see their manuscript published. The number of these publishers (judging from the number of messages in my mailbox) does not seem to be decreasing, thus there must be enough scientists being misled who keep them afloat. Consequently, a few words of advice are in order so that the reader can avoid falling into any such traps.

Because traps they are: if your work is not published but you have paid up, of course you are worse off. However, an article published in such a forum brings you no benefit; actually, your reputation will be tarnished. You should absolutely avoid such journals. But how can you recognise them?

A proper scientific journal relies on declared peer review, and does not try to impress prospective authors with unrealistic claims. Predatory forums (I am reluctant to call them "journals") advertise themselves aggressively, usually through unsolicited e-mail messages, inviting you to become their editors or to submit manuscripts that are, as a rule, promised to be published with extreme speed. Their homepages are unsophisticated, frequently contain glaring errors, display no, or very few, published articles, or you cannot find any content behind apparently "published" articles. Do not rely on their list of editors, because they often list editors' names without their consent. Librarian Jeffrey Beall started a website listing such journals but it was discontinued due to harassment and threats. The scientific community has since revived the list and you can find more information and lists of predatory journals, hijacked journals, predatory publishers and useful related information

Primary and Non-Primary Literature

The first, and most important, matter in publishing is the distinction between primary and non-primary publications. A primary publication is defined by the Council of Biology Editors (CBE, 1994) as:

> [...] the first disclosure containing sufficient information to enable peers to 1) assess observations, 2) to repeat experiments, and 3) to evaluate intellectual processes; moreover, it must be susceptible to sensory perception, essentially permanent, available to the scientific community without restriction, and available for regular screening by one or more of the major recognised secondary services [...].

This definition was carefully considered and merits some explanation. Naturally, the novelty of the scientific information is paramount. More than one publication of the same information is not possible, mainly for ethical and legal reasons. Thus, it also must be a "first disclosure".

The availability criterion has three main components. First is the language. For a long time in Europe, the language of science (as well as culture in general) was Latin. With the development of modern national languages and nation states, this monopoly gradually gave way to different languages. The advancement of the field often spurred the increased use of the language of the country where particular developments took place. Thus, German was an important language of taxonomy and biology in the 19th century, as well as of physics in the first half of the 20th century. Today, the language of international communication in biology (and many other scientific fields) is English. This may put an extra burden on non-native English speakers, but it makes international communication rather simple and (potentially) effective. Practically, papers published in a language other than English may qualify as a primary publication in the case of an intellectual property right dispute but, in "everyday" science, they have rarely been considered as such. One proof of this general perception is the existence of numerous scientific papers published in a less accessible language (usually in a small journal that publishes articles in the language of its country) and also in English, in an international journal. A word of caution, though: journal editors may now consider such articles as attempts to double-publish (Uchmanski, 2019). Perhaps the increased sophistication of automatic translation services is behind this shift — whether this becomes a trend, or the above example remains an oddity, it is too early to tell.

The second availability criterion is circulation. Peer-reviewed journals are much more widely circulated than other publications, such as yearbooks, government publications, and conference proceedings. This is the main reason why papers published in the latter types, irrespective of their quality, do not qualify as primary publications. However, the rapid spread of Internet-based publishing will soon make this criterion obsolete: now even small universities are making full MSc and PhD theses freely available on the Internet via repositories.

It is also important that a primary publication is covered by at least one, and preferably more, of the widely circulated abstracting publications and online scientific databases. Every field has at least one of these, with Index Medicus, Biological Abstracts, Web of Science, or Zoological Records being widely used. These publications make available the bibliographic information of a publication, the addresses of authors, keywords and the abstract. Originally produced in print, they are all now online, and their use has become essential in keeping up with the rapid increase of the scientific literature. As no library can afford to have all, or even most, important primary publications in any field of science within their collections, access to journal articles has become a two-step procedure: a paper spotted in one of the abstracting services or databases can subsequently be looked up and obtained by going to the original journal itself. Finding a paper directly in the original journal by searching the archives of that journal is possible but, due to the sheer number of new papers, rarely happens.

Finally, it is very important that the new information is evaluated by peers, and this evaluation takes place *before publication.*

What Is "Peer-Review"?

Peer review is the process of obtaining written evaluations of a manuscript by experts in the field. On arrival to the editor's office, the manuscript is sent to one or more (usually at least two) experts of the editor's choosing, who will read and comment on it. The peer reviewers should consider the presentation of the problem, its significance, the suitability of the methods chosen, evaluate the reliability and the significance of the results, the views and interpretations expressed in the discussion, the necessity and clarity of the tables and figures as well as the language. The reviewer is required to send the editor a written review that evaluates the above aspects. The reviewer is also requested to make a definite recommendation about the acceptability of the manuscript for publication (usually: accept, reject, or accept with minor/major modifications).

This review is often (but not always) anonymous. This is the "single-blind" system: the reviewer knows who the authors are, but the authors do not know who the reviewer was. A reviewer can always sign the review, but only if she so wishes. When the peer-review system is "double-blind", neither party knows the other: the reviewer does not know the authors' names, because these are removed from the manuscript, and the name of the reviewers are not released to the authors.

The system is not faultless, and various assumed biases have been long debated (Weber et al., 2002; Smith, 2006; Resnik, 2011; Sopinka et al., 2020) but no one has come up with a better system of evaluation that has been widely adopted. The reviewers are the editor's essential helpers, who, collectively, try to make sure that the quality of science published is reliable.

An interesting new system was started in 2011 by Finnish scientists, the "Peerage of Science". They are trying to reinvigorate the peer review system by forcing it to conform to deadlines, a clear track record and mutual assessments (see more at www.peerageofscience.org). The reward for participating reviewers is a record of keeping deadlines, attracting a score of the quality of their review from their peers. Another attempt to document peer-reviewing activity, and to get credit for this, is the Publons initiative (www.publons.com). Others (Fox and Petchey, 2010) have suggested "privatising the commons" where only those who themselves contribute to it can enjoy the benefits of peer review. Yet another attempt is to completely remove the review process, and let the reading public, collectively, decide about quality. Unfortunately, this last system seems doomed simply because of the huge amount of information available, regardless of any other factors that make it attractive or not. The Internet is already the least reliable of information sources (see Chapter 14), precisely because of the lack of systematic quality control of its content.

Non-Primary Publications

There are many types of publications that do not fulfil the criteria of a primary publication. These are not worthless, nor are they necessarily of low quality. Some of these are mentioned below.

Conference Proceedings

Thousands of conferences are organised every year. These can be very useful occasions for discussion, presenting new theories, findings, and so on. Frequently, the conference will have a permanent record in the form of conference proceedings. These are, mostly, occasional publications, containing written versions of varying length, detailing the talks and posters presented at the conference. Conference proceedings are usually published in small print runs, and distributed only to conference participants plus a few libraries that seem to be selected according no particular set of criteria. It is, therefore, uncertain as to whether even the largest scientific libraries will have a copy of the proceedings of any particular conference. They *do not qualify as primary publication*, because they usually fail to fulfil the circulation criterion. More and more such proceedings, however, are electronically published, so one may argue that on the availability criterion alone, they cannot be classified as non-primary. Other problems, however, remain.

One of those is the quality of editing. There are no strict rules about the review process for the inclusion of a specific contribution in a volume of conference proceedings. The editors are usually the main organisers of the conference and, often, do not have editorial experience. Additionally, the conference frequently does not have enough money to pay for the services of professional technical editors, which leads to very variable production quality and publication speed. All too often, such proceedings serve as a permanent record of the conference, a testimony of the organisational work, rather than a vehicle for scientific information—and the thicker the volume, the more important the conference is supposed to have been. This is fallacious logic.

Occasionally, conference proceedings are published as special issues of a scientific journal — in which case they are subjected to the same review process as any other manuscript submitted to that journal. Not all contributions presented at the conference will necessarily be included, but the ones that satisfy the process should be considered as valid primary publications. In any other case, do not consider conference proceedings as primary publications. This does not mean that you should never write one — for advice, see Chapter 27.

Government and Institutional Reports

These are usually obligatory reports written on work done using government grants, expert committees, or an output of commissioned research. Their primary purpose is to inform the funding organisation about the work carried out, often concentrating on the recommendations emerging from the project. They rarely qualify as primary publications because they are seldom peer-reviewed, have limited circulation, and are often in a local language.

Review Publications

A special set of non-primary publications are the review journals. These exclusively publish reviews of published literature in a specific area. Because reviews are the first source of synthetic information when one wants to get quick orientation in a new field, review journals are read by many people, and can be very influential. An additional reason for their popularity is that, with the increase of the primary literature, more and more people find it difficult to follow the developments from reading the primary literature only. Review publications (Box 1) can be yearbooks such as the Annual Review series, and even monthly, like the journals in the "Trends" (Trends in Biochemistry, in Ecology and Evolution, in Genetics, etc.) series.

Box 1. The most common review journals

Annual Reviews

A California (USA) — based non-profit organisation, publishing annual volumes of authoritative reviews in many fields of science, the current range includes 45 Annual Reviews. Their articles are usually the most cited in their respective fields.

The "Trends in..." journals

A series of monthly journals, started by the international publisher, Elsevier, Netherlands. Now published by Cell Press, they include 14 journals, published monthly, in a range of areas across the biological sciences. They publish short, topical reviews, correspondence, comments, book reviews, etc. The reviews are less substantial, but it is a useful source of keeping abreast of new discussion topics in a field.

Critical Reviews in...

A series of 13 journals, published by Francis & Taylor, a UK-based publisher. These cover areas from chemistry to philosophy. See:

Current Opinion in...

Elsevier's Current Opinion journals comprise of 13 titles in life sciences and adjacent fields.

Reference Literature

There are only estimates of how many scientific journals are published today (about 55,000, Gastel and Day, 2016). There is no library where all these can be found. The need to be informed about the availability of a new piece of scientific knowledge is greatly met by the existence of reference journals or publications. These publish only selected parts (sometimes only the reference) or summaries of primary scientific articles. The idea is to provide some extremely condensed information about the article, and its necessary bibliographic data, so that interested readers can find and read the original, full-length version. Many of these, until about the late 1980s, were produced by teams of reviewers who received, read, and abstracted the original papers. These "abstracting" journals have titles such as *Biological Abstracts*, *Plant Protection Abstracts*, etc., and contain summaries of varying lengths of selected original, longer papers, plus assorted indices to help interested readers to find the sources if needed. Today, probably all of them are produced by at least semi-automatic means, containing the (usually unaltered) abstracts of the papers, and various computer-generated indices of topics, authors, and affiliations.

The most successful of these, originally named *Current Contents* (which became Web of Science), and published by the Institute for Scientific Information in Philadelphia, USA, also gave rise to the science of scientometrics.

By a stroke of genius, the founder of *Current Contents*, Eugene Garfield (1925-2017), abandoned the manual extraction of information from articles, and only published the photographed contents pages of journals, plus indices of topics, authors, and addresses. Anyone reading the title of an article, if she was interested, could find the contact details of the author, and could ask for a full copy of the original article. The disadvantage was that a reader had to decide, after reading the title only, whether the article in question contained relevant information. However, the advantage was huge: while normal abstracting journals usually published their more detailed, but still partial, information about published papers with a delay of 6 months to -2 years following publication, *Current Contents* published the (admittedly more limited) information in 6 weeks — to 6 months. This was possible because most of the work could be done much faster: it only involved photographing and printing the contents pages, and generating the relevant indices. This time-saving method proved the definitive factor in the success of *Current Contents*.

Current Contents was organised in sections, and the original, selected range of journals has gradually expanded to include, today, >18,500 scientific journals. The latest development has extended to the listing of conference proceedings and books. There is still a bias towards US-based scientific journals published in English. When the Canadian publishing giant Thomson Reuters became the owner, several doubtful practices emerged as they aggressively, and rather inconsiderately, extended the claims for what such a database can achieve. Now this database (Web of Science, or Web of Knowledge) is no longer part of the Thomson-Reuters media empire and, despite vigorous protests from academia (see www.sfdora.org), the current website still claims that it offers "an objective analysis of people, programs and peers".

3

Citations and Scientometric Indices

While the quality of science is extremely difficult to measure, the field of scientometrics attempts to do so by studying how the impact of scientific publications can be measured. The task remains elusive, but one system, quite widely in use now, argues as follows.

In a scientific paper, there are only two types of factual statements: already published, known information, which is necessary to enable people to understand how new information relates to earlier material, and the new information. While the new information is supported by the facts, data, figures and tables presented in the paper, the known facts are simply mentioned, with the reference to a publication where the relevant fact was first proven/published. This is called a citation. The precise bibliographic data of such citations are listed at the end of published papers, and they can be identified, counted, and summarised.

Important findings, goes the argument, generate new research, and when the new discoveries are published, these previously published findings are cited as connecting links to the understanding of the new discovery. Such papers are therefore frequently cited. This approach equates high citation rates with high "impact", which, according to this simplified perception, also indicates high importance and/or quality.

It is easy to see that, even if we accept the above argument, a few key questions must be decided: what counts as a citation, where do we do the counting, who does the counting, and for how long?

This is where business sense and sharp thinking came together to create a business opportunity, as well as a new field of analysis. Using its unique position, the Institute of Scientific Information (abbreviated to ISI; but one should not be misled by the name — this was not an institute, it was a business venture, publishing *Current Contents*) declared that a) we — ISI — will do the counting; b) a citation counts only if it appears in a journal covered by our publication, *Current Contents*, and; c) citations are "valid" and counted over a period of only 2 years after the publication date.

Originally, the purpose was to identify the most influential *journals*, and according to the ISI philosophy, these were journals that published the most frequently cited articles. Citation (only during the 2 years after publication, remember) equalled scientific impact, and the index thus coined was named the "impact factor" (abbreviated to IF). Despite discussions and doubts almost from the beginning, IF has caught on and, today, there hardly is a scientist unaware of the term. The success of *Current Contents* had a knock-on effect on journals, and the ones with a higher IF had an advantage over their rivals, in terms of distribution, recognition, and competition for manuscripts presenting discoveries that were thought important. The same statistics were soon applied to organisations and even to individual scientists, and when ISI was sold to Thomson-Reuters, aggressive promotion of these more dubious uses intensified.

A multitude of indices based on citation statistics has appeared since this original index, and there are several books and fora discussing their merits and demerits — the reader is directed to some of these; as a first step, to the ISI website itself, which today calls itself "Web of Knowledge" (https://www.webofknowledge.com). Here, only two of the most widely known indices are mentioned: the impact factor (IF) and the Hirsch index (*h*-index).

The IF of a journal is defined as the average number of citations that a single article, published in that journal, receives in the range of journals covered by Web of Science in the two years after publication (see Box 2 for an example of how to calculate IF). It is worth pointing out — even if this has been done many times — the hubris that the naming of the statistics displays. Being a competitive species, humans could not resist taking the next step, from ranking journals this way to ranking scientists following a similar logic: scientists who publish in high-IF journals are important scientists, and those who do not, are not. There are many pitfalls along that route, and for a more detailed discussion, readers can find several sources; a good recent example is Mingers and Leyesdorf (forthcoming).

Even if we accept the above logic for assessing individual scientists, the use of the IF to rank journals where one publishes is imperfect — the IF values are averages, while the distribution of citations are very right-skewed: very few articles get much more than their expected share of citations, and become fashionable, or "citation classics". Most articles get much less than the expected average number of citations: thus, the overall distribution of citations is very right-skewed. This was named the "Matthew Principle", a tongue-in-cheek reference to a passage in the Bible (Matthew 25:29, RSV) claiming that to those who have, more will be given, and the poor will lose even what little they have.

Given this state of affairs, a second, more logical, step was to use the number of actual, rather than potential, citations to assess scientists. Again, a multitude of indices have been suggested (Harzing, 2002); currently, much in vogue is the Hirsch-index, or *h*-index (Hirsch, 2006). To calculate someone's *h*-index, all her publications are ranked according to the number of citations attracted, from the highest to the lowest. A person's Hirsch index equals the number where the number of citations for any individual paper is not smaller than its rank number (see Box 3 for a calculated example). Several modifications and alternatives have been suggested, and the reader can find a good summary of these in the help files of the program "Publish or Perish", developed by Anne-Wil Harzing (see her website: www.harzing.com).

To be included among the journals covered in *Current Contents* originally, a candidate journal had to fulfil stringent criteria: regular publication according to a schedule, papers written by an international range of authors and on topics of wide interest, and a reasonably wide international distribution. Journals usually must wait for at least three years before they can get their first impact factor. Journals are now also ranked by their relative position in their category (occasionally in several categories), usually by quartiles (e.g. a Q1 journal is in the top 25% of its group); sometimes the top 10% also forms a separate class (called D1).

Box 2. The Impact Factor (IF) and its calculation

The so-called "impact factor" is, in fact, a very limited index, a combination of hubris, business acumen, impatience, and disregard for the genuinely novel. It was introduced by the then-Institute of Scientific Information, a business venture (not a research institute, only in the name) publishing the shortcut-to-scientific literature, Current Contents. The Impact Factor was defined to compare scientific journals, and is the average number of citations a scientific paper published in a journal receives in articles published in the journals appearing in the former Current Contents, now called Web of Knowledge (also known as Web of Science, WoS) in the two years after publication. Now also the 5-year impact factor is published, which is the number of citations in the same journals in the five years after the year of publication:

Journal IF year x= no. of citations in WoS journals, in year x, to articles published by journal in year x-2 + in year x-1 / no. of articles published by journal in years x-2 + no. articles published in year x-1

The journal Urban Ecosystems in 2014 has an impact factor of 2.685. It is calculated the following way:

Citations in 2014 to items published in:	2013 = 96	Number of items published in:	2013 =51
	2012 = 202		2012 = 60
	Sum: 298		Sum: 111

$$\text{Impact factor} = \frac{\text{citations to recent items}}{\text{no. of recent items}} = \frac{298}{111} = \mathbf{2.685}$$

Box 3. How to calculate the Hirsch-index, *h*

This index was suggested by E.J. Hirsch (2005).

In order to calculate the index, first one has to rank all one's publications according to the number of citations it received, in a descending order.

The index is the value of the rank where the number of citations received by that paper is still larger than its rank number. It is claimed that this index does not depend on the publication activity in a field. This claim is clearly suspect, because the number of citations depends on the size of the "citing universe" as well as the citable universe (the number of one's own publications).

The Hirsch-index tends to favour senior academics, because they can have a longer list of papers. It also underemphasises the highly influential papers. In order to achieve a Hirsch-index of 20, it does not matter if the top-ranked article is cited 20 times or 200 times. For example, one can have an *h*-index of 20 with 20 papers and 400 citations — if all 20 were cited 20 times — and someone with 4000 citations can have a similar h=20 if her 21st article in the rank was cited <21 times.

	Number of citations to paper by rank						Total no. of citations	Hirsch-index
Name	1st	2nd	3rd	4th	5th	6th		
AA	20	10	5	4	4	4	47	4
BB	150	120	100	92	4	4	470	4
CC	5	5	5	5	5	5	30	5

As the examples show, the total number of citations does not matter, only the number of citations of a paper relative to its rank — thus, AA and BB have equal *h*-index values. CC only has 30 citations, yet his *h*-index is the highest of the three. Nevertheless, this index, due to its simplicity, has become a current favourite, especially among science administrators.

It is suggested that a Hirsch-index of h>20 indicates internationally significant scientific output, and an h>30 an exceptionally influential one.

There are several related indices that aim to correct the identified disadvantages of the Hirsch-index, such as the age of the publication, the number of authors, or the size of citation "excess". A good summary can be found in the help pages of the Publish or Perish software or the accompanying book (Harzing 2010).

The citation statistics of thousands of journals are collated and published in the *Journal Citation Reports* (JCR), issued yearly by Web of Science. These statistics, available only by subscription, are widely known, popularised, and used for various purposes. Recently, a few alternatives have emerged. Scopus (www.scopus.com) collects citations and various scientometric indices from the Internet, but its coverage of the literature is limited. This is a for-payment service, but the freely available program "Publish or Perish" (see above), calculates numerous citation statistics, using information in the free database Google Scholar. Harzing runs a well-maintained website, and published a book (Harzing, 2010) that describes many of the advantages and disadvantages of using scientometric indices. Google Scholar itself also has the capacity to calculate scientometric indices that can be used by any registered visitor. Both platforms are less English-biased than Web of Science.

Citations have become the dominant way of measuring scientific impact, and various statistics related to them are followed, counted, collected, documented and used by scientists themselves, as well as by journals and various science-related organisations. Citations are also being manipulated in various ways, the easiest of which is self-citation. This is done by journals as well as individual scientists and consequently, today, there is a distinction between "independent" and "dependent" citations. A citation counts as independent if no author of the citing document is an author on the cited document. If even one of the cited authors is also a citing author, this is counted as dependent or self-citation.

In general, there is much to resent in the superficial use of scientometric indices, and scientists must engage with science administrators to increase the mutual understanding of the benefits and limits of these methods. I suggest that readers familiarise themselves with the basics of scientometrics and become aware of some of the major controversies, because the use of such statistics is not going to disappear from science. The field is fast developing, with a major academic journal, *Scientometrics*, and numerous books (e.g. Vinkler, 2010) dedicated to the topic. The misuse of scientometrics lead to the San Francisco Declaration that provides guidance to the various parties engaged in science, from practice to policy.

4

Things to do before Beginning the Process of Writing

Here, I begin with the first principle of communication: who is your expected readership? Writing is made easier when one knows the exact goal, so it is best to make some basic decisions before sitting down to write. The most important question is: what type of article do you want to write?

There are obvious differences between a short communication, a full-length primary article, a review, a book chapter, or a thesis. All have their (rigidly enforced) rules about structure and format that must be observed before a manuscript is even considered for publication. Additionally, even though there is a general structure to a scientific paper, various journals follow different formats, and these must also be adopted before an editor will subject a submitted manuscript to an assessment of its scientific quality.

A second, equally important question: who is the intended readership? Who do you want to communicate to? Are they specialists, generalists, or lay readers? Do you expect them to be colleagues in your home country, or is the expected readership more international, perhaps even from different fields? This should strongly influence the level of detail and style of the paper. In order to use the appropriate language (coding) when describing the new information, it is very important to consider the readership. Recall the general principle: the simpler the code, the easier is the decoding (or understanding).

How to Decide Where to Send a Manuscript for Publication?

A devil's advocate might claim that it is not even worth asking this question. Why would this possibly be important in the age of extensive literature databases? Would it not be easy for everyone to find a paper, irrespective of its publication forum, using the widely available Internet-based search engines? Against this advice, I argue that it is still very important that papers are published in the most appropriate forum. The wrong choice of journal can result in one of the following things, none of them pleasant to the author:

a) Rejection without review due to the paper being "not in our scope".

All journals have a defined scope, identifying the area in which they aim to publish scientific papers. Manuscripts that, in the judgement of the editor, fall outside the scope of the journal are simply sent back to the author as "not suitable for our journal". This causes needless delay for the authors who are usually anxious to see their paper published as quickly as possible. They also have the additional frustration to having to re-format the manuscript before it can be sent to another journal.

b) Inappropriate review.

Another potential unwanted consequence is that, while the journal editor decides that the journal could potentially publish the paper, the topic of the manuscript is not really in the mainstream area of that journal. Consequently, the journal may not have expert reviewers in the field, and the work gets an unfair review, simply because the reviewers are not familiar with its area. A common human fault is that, if the reviewer does not understand the work, the manuscript gets the blame as incorrect or badly written. Rarely, the opposite might occur and the manuscript may be accepted even though it is faulty; more frequently, it is rejected due to unjustified criticism. Several months may pass until this becomes clear, and the authors have again lost precious time. Further, even unjustified criticism hurts, and authors would do well to avoid it.

c) Publication without effect.

Even if the manuscript is accepted and published, it may turn out that papers on this topic are so rare in that journal that people working in the relevant field have stopped regularly checking it, and thus few colleagues would see or read the paper. The publication may thus "sink without effect". If you hope that your peers will pick the paper up anyway when it appears in the secondary (review) literature, you must consider that secondary review services usually pick the original papers up only after a delay. Moreover, do not overestimate the efficiency nor the frequency by which your peers search and read secondary literature.

Therefore, aspiring authors would do very well to carefully target their manuscript at a specific journal. Before choosing a potential forum for your manuscript, it is worth consulting:

- *Your colleagues*. Most scientists are keenly aware of the major journals in their field and can give good advice of their scope and practice. They may offer you their practical experience of the journal you are considering.
- *Secondary review journals*. These often publish their sources grouped by fields. Scanning these would indicate to any aspiring author a range of potential journal choices.
- *Mastheads of the possible journals*. This usually appears on the inside cover, stating, among other things, the purpose of the journal, the types of articles it accepts, and a host of other useful information.
- *Instructions to authors*. This section often elaborates further on scope, types of communication published in the journal, preferred or acceptable types of manuscripts, and any limitations by geography, topic, length, etc. Instructions to authors are always freely available on the journal website, and are published at least once a year in the printed journal.
- *Recent issues*. Consulting actual issues is useful, because they show how the current editorial team interprets the mandate of the journal, and what kinds of papers really do get published. Editors usually serve a finite term, and every editor will interpret the task in slightly different ways, putting emphasis on different fields, types of papers, etc. Consulting recent lists of contents can give

useful information about this — important but often unspecified — aspect of editorial policy.

Further, it is also worth checking:

- *Publication schedule.* A journal that is published more frequently will have a faster turnover, offering the possibility that a paper is published earlier.
- *Actual publication dates.* All journals have a publication schedule, but not all of them can keep to it. If, by March 2005, the November 2004 issue is still not available, there is something wrong with the journal.
- *Handling time, printing time.* Authors would like to see their work printed as soon as possible. However, the publication process can take several months. With a little calculation, aspiring authors can find out about the length of this period if they can find a number of dates. The first is the date of submission. It is worth knowing that this is also the "official date of discovery". The date of acceptance is also usually printed. The difference between the two dates, calculated for several articles, indicates how fast (or slow) the manuscript evaluation process typically is in that journal. In biological journals, this can be 2-8 months. The time difference between the date of acceptance and the date of publication (to be found on the cover of the issue), indicates the time necessary to turn an accepted manuscript into a printed paper. It rarely takes less than 6 months, but electronic publishing is usually shorter (2-4 months). If this information is not available, treat that as a warning signal. All journals strive to be fast and, if they succeed, they certainly "publicise" it by printing the above dates. If they fail, they may not advertise their failure — and only print the date of acceptance.
- *Publication standard/quality.* The general appearance of the journal is also a useful guide. Carelessly prepared journals, with bad quality figures and printing, often indicate that the journal's standards are not high, or the journal is not financially secure (and is not able to afford a better, costlier,

printer). Such journals are best avoided after all, you would not want your wonderful work to be ruined by careless, ugly printing, would you? Additionally, in some cases, you may have pictorial evidence, for which the quality of printing may be crucial. Bad printing can ruin evidence.

- *Circulation.* The wider the better. A journal in which all papers are Open Access will circulate your results to a larger audience. For others, one can only guess: from the affiliations of the editors, the reputation of the publisher, the affiliation of the authors, or the study locations. An international journal should not only have editors from one country, nor will all the studies come from one country or region. Look, also, for biases — many journals try to sell widely, yet they may, mostly, publish articles from one region (e.g. North America or Europe). Manuscripts from other regions may have more difficulty in getting accepted.
- *Cost of publishing.* Publishing in European journals is usually free (except for printed colour figures, see later), but many journals published in the USA, Canada, Australia and New Zealand have a system of page charges. In such journals, authors are expected to contribute to the cost of publishing their papers. Authors cannot buy the right to publish, and, in many cases, your ability to pay page charges does not influence your chances of getting published. Many journals, though, assume that by submitting your manuscript, you will be able and willing to pay publication costs in the case of acceptance. These charges are based on a printed page, and can vary from US$30/page to US$1000/page. Sometimes, members of a society can publish for free, or at reduced cost, in the society's journal. Open access journals routinely require that the authors pay for all the processing and publication costs (article processing charge). If you do not have funds to pay for these charges, it is not impossible to publish in those forums, but your inability to pay at all, or ability to pay only partial costs, should be clearly indicated in the accompanying letter when first submitting the manuscript. The editor often has discretion over page

charges, and can allow you to publish without paying page charges, or lower the charges. The important thing is to be open and honest and indicate your inability to pay at the time of submission.

- *Access to your paper.* Once published, readers will have access to your paper in various ways. If your paper is published in an open access journal, everyone can freely access it. The same applies if you, in a subscription-only journal, paid an extra fee to make *your* paper open access (this option is quite expensive, though, and is known as "hybrid" open access). If the journal appears in print, you may get reprints, copies of your paper only, printed and stapled separately. Most journals will give 25-50 reprints for free; more can be ordered at the proof stage (see more about reprints later, in the chapter about proofreading (Chapter 25)). Reprints, however, are nearly extinct — today authors will more frequently be given electronic (usually pdf) files. As a rule, these can be sent to other individuals. The website https://www.howcanishareit.com/ can also be consulted for getting advice how can you share your article.

Most of the above information is available on the journal homepage. Such homepages contain at least the following:

- Editor's name and (usually postal) address.
- Editorial Board members' list, usually with (postal) addresses.
- Frequency of publication.
- Information on aims and scope.
- Detailed instructions to authors
- Page charges — if any — and possible exceptions.
- Information about authors' rights to use/distribute their own paper, including the number of reprints they get for free, and any eventual restrictions for on the author's own use.
- Addresses for correspondence (usually includes e-mail).

Additionally, the content pages are always freely available and, also, in most journals, the abstracts. Even if you have no subscription to a journal, you can still check at least some papers. Journals may run promotions, when a selection of full papers is temporarily available for free. Some — otherwise not open access — journals have a policy of making their full contents freely available after a set period (for example, all papers in the *Proceedings of the National Academy of Sciences of the USA* are freely available after 6 months of publication). In such journals, many papers are freely available, because the authors have paid an extra fee for open access. These papers are indicated on the page of contents and these can be downloaded and printed for free. You do well to consult at least a few full papers — some of the above information can only be found within them.

Collecting the above information on a range of journals is made easier by consulting the Internet (e.g. see Box 4) and it takes some time on the first occasion, but you do not always have to start from scratch. Once you become a publishing scientist, you will soon develop experience of, and a feel for, these aspects. A little time invested at this stage will save you a lot more time when it comes to submission and publishing.

Box 4. Sample Internet addresses of selected journals in the field of biology

Ecology and Society: www.ecologyandsociety.org

American Naturalist:

Oecologia, Behavioural Ecology & Sociobiology, Marine Ecology:

Cambridge University Press journals:

In this section, the different elements of the primary paper will be discussed, in the sequence in which they normally appear *in the manuscript*. The structure of a primary scientific paper is sometimes abbreviated to "IMRaD". This means "**I**ntroduction, **M**aterial and methods, **R**esults **a**nd **D**iscussion". There are, however, several additional parts of a paper (see Box 5) that appear in a specific sequence in the manuscript. This sequence is *different* from the order in which they will appear in the printed paper. The most important difference is the placement of the figures and tables. These should always be placed at the end of the manuscript, and not pasted into the text. The original reason for this is that the printing of figures was a separate process, carried out at a different place to the printing of text. Even today, illustrations are processed differently from text (often using different software, too) so inserting photos into the text of the manuscript would serve little purpose. Tables and figures should always be at the end of the manuscript. More detail on this matter will be given in the respective chapters on figures and tables.

Box 5. Parts of the manuscript of an article

Title

Authors =>

Addresses =>

Corresponding author and her address: =>

========================> the above often form the *title page*

Abstract/summary (sometimes also in the form of "highlights" and "graphical abstract")

Keywords

Short title/ running title

Introduction

(continuous...)

Material & methods (this is sometimes placed to the end of the paper)

Results

Discussion

Acknowledgements

References/literature cited

Tables

Figure legends

Figures

Appendices (optional)

In a few journals, the material and methods section is placed at the end of the article. This, the editors argue, makes reading and understanding easier, because the background (introduction) is closely followed by the new results and their interpretation (the discussion). Irrespective of your opinion about this, you must follow the convention in the target journal.

PART II

The Process of Writing the Paper

5

Title and its Composition

Things should be made as simple as necessary but not more so.
Einstein

Why Is the Title Important?

The title is, arguably, one of the most important parts of a paper. The title is not only printed first (or very close to the top) in an article; it is the part that will be read by most people. These readers must then decide whether the paper contains information that is relevant or interesting for them. The main requirements of a good title can be formulated briefly and sharply: be precise, simple, and short. The best title is one that gives the most accurate information about the content of the paper with the fewest possible words.

When formulating the title, one should also consider that many potential readers will only see the title in the different abstracting journals and services, and this constitutes the information on which they will have to decide whether to obtain a copy of the paper. Abstracting and information retrieving services, as well as Internet search engines, also use the words in the titles.

Loose or imprecise words in the title generate inexact search results. The rules that readers follow are very simple: they will pass over loose, ill-defined, or overly general titles. They will not be intrigued by cryptic titles, jokes, or vague promises. They will know that there cannot be a *"Theory of everything"* (Laughlin and Pines, 2000). Having an informative title is mandatory, otherwise the paper will never reach the intended readership.

In earlier literature, overly general and imprecise titles occur often. The results presented in a paper in 1957 under the title, *"The influence of some cations on an adenosine triphosphatase from peripheral nerves"*

(Skou, 1957) eventually earned its author a Nobel Prize. However, it is impossible to judge from the title what was novel in the paper, or even what was studied precisely. It would certainly not be considered an effective title today. The Nobel Prize was probably not given for the excellence of the title!

Developing an Effective Title

Generally, the title should indicate answers to the important basic questions: *What? Where? How?* What was studied: an organism, a mechanism, a community, a molecule? What was measured? And what methods were used? Where: in the laboratory, or in the field? How was the study organism or phenomenon examined? A good title indicates all these elements, so that readers can judge what to expect in the paper.

Our hypothetical manuscript contains the results of an experiment studying the effect of root exudates of 15 cruciferous plant species on food consumption by one species of caterpillar, *Helicoverpa armigera*, an important pest world-wide. Caterpillars were kept on plants to which five different concentrations of root exudates were applied, and the caterpillars' growth (body mass) was measured over their larval period.

Firstly, consider the title *"The effect of chemicals of plant origin on caterpillars"*. This title is short, but it is misleadingly general. Literally, this would mean something like examining the effects of ALL compounds that can be isolated from ALL plants, on ALL species of caterpillar, using ALL possible reaction parameters, ranging from individual behaviour to mortality. It does not say if the study was done in the laboratory or under field conditions. Consequently, this would make a very poor title.

Would it be more precise if all 15 plant species were listed? Yes, but this would make the title impossibly clumsy and long. This would go against the requirements of brevity, so it is not a good solution.

How about *"The effects of 15 plant species on the larvae of* <u>Helicoverpa armigera</u>"? This is an improvement, because it indicates the number of plant species and the insect species on which the study was performed. However, important information is still missing: *which* 15 plant species were used in the study? Can they be specified somehow? Exactly *what* was measured on the caterpillars? *How* was the "effect of plant species" generated?

In order to answer the above questions, the title could be modified to "The effect of root exudates of 15 cruciferous plant species on the growth of *Helicoverpa armigera* larvae", or even "Inhibition of growth in *Helicoverpa armigera* larvae by root exudates of 15 cruciferous plant species". Either of these two examples is now an acceptable title: they indicate the effective agent (root exudates), the range of plants (15 cruciferous species), the reference parameter studied (growth) and the target organism (*Helicoverpa armigera* larvae). It still does not indicate if this is a study carried out in the laboratory or the field. This would be more important if the work reported here were a field study — then, even the location might be interesting.

Series Titles, Hanging Titles, Questions and Statements

Series titles are not usually accepted. The reasons for this are, at least in part, practical: there is no guarantee that, because 13 previous articles have already been published, no. 14 will also be accepted for publication. All manuscripts are assessed solely on their own merits. Second, manuscripts are processed at different speeds, even in the same journal, not to mention different journals. For example, if manuscripts no. 13 and no. 14 were submitted to different journals, there could be no guarantee that no. 13 would be published before no. 14 (or published at all). Of course, in such a situation, you cannot ask to delay the publication of no. 14 until no. 13 is published. Such series titles serve very little purpose, apart from advertising one's monumental achievements such as a twenty-paper series. Most journals will be unwilling to print such a series as it adds little to the specific scientific problem studied in the papers. It is best not even to try.

It is a recent fashion to have hanging titles. The title "*Doing it right: the art and science of publishing*" would be a (poor) hanging title for this book, for example. As is often the case for such titles, the first part is not necessary. Many journals now seem to accept, or even promote, the use of hanging titles but many authors see this as an excuse to give two titles. Try to avoid this. Sharp thinking and brevity is always better.

The title is not the appropriate place, either, to advertise your knowledge of pop culture, sense of fun, etc. These elements are often unnecessary, do not add to the precision of the title and, thus, do not help the reader. What sounds funny at a party is certainly less so when printed in 14-point bold letters at the top of your most important discovery.

Also, be aware that the readership may (and, you hope, will) be drawn from a very wide cultural and linguistic circle, and they could be baffled by several "subtle" references. They may even be offended and that should be avoided.

Questions do not usually make good titles. In most cases, the readers are not interested how fascinating questions can be formulated — they want answers. Occasionally, a question can be a powerful title — but only exceptionally. For example, if the research question cannot be answered in a simple way, you can formulate a title that contains the question asked: "*Does urbanisation decrease diversity in ground beetle assemblages?*" (Magura et al., 2010)

In some cases, the main results of the investigation can be summarised in one sentence. Such a statement can form an effective title, such as: "*Insect feeding mobilizes a unique plant defense protease that disrupts the peritrophic matrix of caterpillars*" (Pechan et al., 2002). Generally, however, such one-statement titles cannot be formulated and should be avoided. Most studies are more complex and condensing them into one sentence is difficult or impossible.

Syntax and Jargon

The desire to be concise does not supersede the requirements of grammar. In the drive for brevity, look carefully at your syntax and use of jargon. It is not the same thing to suggest a "*New colour standard for ornithology*" or a "*New colour standard for ornithologists*". The former title promises a new tool for the science of ornithology, while the latter may be helpful if the aim is to distinguish ornithologists of different colours, a somewhat improbable purpose.

Abbreviations and jargon, chemical formulas, trade names, and similar words should also be avoided in the title. In case of doubt, it often helps to ask the question "would anyone find this word if searching in an index?" If possible, unusual words and other non-general terminology should also be avoided.

Long titles are often too long because they contain non-specific "waste" words, or "fillers". Such words often occur at the beginning of the title. "Studies on", "Observations on" "Description of" and "Investigations on" are typical examples of such fillers. Starting the title with *The, An* or *A* is also a waste because these are also uninformative for the indexing services. They are sometimes necessary for correct grammar, but consider carefully if they can be avoided. Similarly, words such as "changes", "effects", "impact", or "trends" are not powerful. Changes are everywhere in nature, so, in detecting a change yourself, you may not have discovered anything novel. Similarly, experiments are designed to detect effects or impact — no need to advertise this in the title. *What* kind of effect, a change in which direction — indicating this would be more informative.

When to Write the Title?

The writing process is rarely "linear": very few people can write a manuscript from the beginning to the end in one go. If you, my reader, are one of those fortunate people, you probably do not need this book. I envy you. Most of us, however, are not like this: we do not write in the same sequence as we read. Even though it will be printed (and read) first, you do not have to have a perfect title before moving on to writing the other parts of a manuscript. Formulating a title can be left to a late stage of the writing process.

My suggestion is to spend a little time at an early stage in the writing process, and jot down a few key words that you feel should be in the title. This will be your *provisional title*. After this stage, the title can be put aside, and you can start working on other parts of the manuscript. At about one third of the way into the writing process, return to the title. Now you will have a clearer idea of how long the paper will be, what the focus will be, and which aspects will be emphasised. Keeping these points, and your intended audience, in mind, try to formulate a more complete *working title*. Finally, near the completion of the manuscript, when virtually all the writing is done, consider the title again, and decide on the *final title*.

Running Title

Many journals print, at the top of each page of the paper, a short title called the "running title". This is a specially shortened version of the title (because of space limitations). If the journal prints such running titles, the specifications — usually in terms of the maximum number of letters and spaces — are given in the "Instructions to authors". It is best to follow this advice, and construct a running title if required. If the journal prints running titles and you do not supply one, someone, usually the editor, will construct one and you may not be happy with the result. After all, who is better qualified to create a meaningful short title than you, the author?

6

Challenges Regarding Authorship

An author, by definition, is a person who brought the work into existence. Given the importance of the first publication of new scientific results, authors of important papers (in other words, discoverers of notable new facts) gain respect in the eyes of their peers and indeed (although not always) the wider world. Scientific publications constitute what matters in science, and thus it is not surprising that scientists care a lot about authorship.

We suggest that any aspiring author should tackle the question of authorship as early as practicable in the publication writing process.

Co-authorship is almost inevitable today; a scientist working alone is a rarity. The average number of authors of a paper in biology was about 2.4 in the 1980s, 4.5 in 2000 and by 2017 it had grown to 6.2 (Kelly, 2018). Publishing as a sole author rarely occurs in a scientist's career.

Several authorship ranking systems are in use even today, but the view that the first author should be the one who did most of the work that led to the paper is gradually gaining prominence. In some fields, or in groups lead by very strong personalities, the situation might still be different, but the trend to list authors according to their contribution to the paper is gaining ground: the first author should have done most of the work, followed by others who were actively involved. Sometimes the first author is called the "senior author" - — perhaps a remnant of the times when seniority indeed decided the authorship sequence.

This principle also means that supervisors or group leaders (managers) should not be automatically credited with authorship on papers. This is a tricky issue because of the kudos associated with authorship — people in power are often unwilling to give up this "privilege". They rarely contribute much, but they use their power, openly or not, to be credited with authorship on papers emerging from "their" laboratories. The argument mentioned most frequently in these cases is that "otherwise nothing would have happened", the opportunity to do the work would not have occurred and, thus, the results could not have been achieved. Following that logic, you might include your parents as co-authors as, very obviously, without them, "nothing would have happened", either. The real involvement becomes obvious when it comes to light that one team member did something wrong, or, worse, falsified data. In those cases, it quickly turns out that the boss "was not really involved in that particular paper". No more needs to be said on this matter.

However, this does not make the matter of authorship, and authors' rank, an easy one. Just as with many aspects of group activity, giving credit to one's own work is a matter of personal judgement. One tends to overestimate the importance of one's own contribution. Conflicts usually arise not when someone gets undeserved credit for something, but when one does not get, in that person's view, the recognition they should receive. Many long-running co-operations and partnerships have broken up due to neglecting this aspect of the publication process. As is often the case with human conflicts, the root of the problem often lies in assuming things and not discussing them. The remedy is simple: openly discuss this issue. Expectations can only be met when they are known. Discussing authorship early will generally smooth relationships and ease co-operations.

This does not mean that the circle of authors and their rank on future papers should be decided even before the work has started. The team, however, can agree on certain principles, or rules, that are accepted by all. For example, the plant ecology unit at the University of Sheffield, UK, published their co/authorship scoring system (Hunt, 1986), which divided the different phases of the work leading to a paper, and assessed individual contributions to each one of these, allocating points to contributing individuals.

Box 6. The co-authorship scoring system used by the plant ecology group at the University of Sheffield, UK

1. Intellectual input (planning/designing/interpreting)

no contribution	0
one detailed discussion	5
several detailed discussions	10
correspondence or longer meetings	15
substantial	20
closest possible involvement	25

2. Practical input: data capture (setting-up, recording, observing/ abstracting)

none	0
small	5
moderate indirect	10
moderate, direct	15
major indirect	20
major direct	25

3. Practical input after data capture: data processing/ organising — but not interpreting see 1.

no	0
minor or brief assistance	5
substantial or prolonged	10

4. Specialist input from related fields

none	0
brief or routine advice	5
specially tailored assistance	10
whole basis of approach (but advice only	15

5. Literary input (contribution to first complete draft of Ms)

none	0
edited others' material	5
contributed small sections	10
contributed moderate sections	15
contributed majority	20
contributed virtually all	25

The group requires a minimum sum of 25 points to become an author, and authorship sequence is decided by the number of points. insufficient number of points are taken over to the next paper — i.e. a colleague who does not accumulate enough points to become an author on a paper, has a "head start" at the next one, as the accumulated points are credited for the new one.

When a paper is ready for submission, all contributors are scored following the system in Box 6. Anyone with a score of over 25 gets co-authorship. The sequence is according to score rankings; scores below 25 are carried over to a subsequent paper.

Authors can also be listed in alphabetical order if no sequence is desired, or authorship can be decided by the toss of a coin. In these cases, this fact is usually mentioned in a footnote on the first page of the manuscript. If two or more authors contributed equally to the paper, this can also be mentioned in a footnote.

The main advantage of developing an authorship sequence decision system is transparency. Everyone in the team knows the criteria, and this channels otherwise potentially disruptive conflicts onto a manageable path. Potential conflicts are not eliminated, but the procedure provides a structured way to handle, discuss, and resolve them. Such a system also places the authorship criteria firmly into the domain of the work done. It is strongly advocated that research groups develop their own authorship decision system. The potential benefits are significant.

The trend "authorship equals real contribution" is reinforced by the recent requirement of having to specify, in detail, each author's contribution to the paper. Several journals (for example *PLOS One*) have developed detailed criteria for authorship.

The expected combined effect of these developments is that, more and more, authorship will reflect real contributions, and not power relations in science. The suggestion that teams should identify the mechanism for how authorship was decided (Tscharntke et al., 2007) seems a sane one and it, at least, provides some information about allocating authorship. However, it does little to clarify principles, nor does it move the field towards the desirable status of preventing colleagues in power abusing their influence to gain authorship.

What Does Co-Authorship Mean?

Co-authors may only have contributed to certain parts of the work on the paper, but all authors bear collective responsibility for the total content of the paper. This must usually be declared at submission. If you are a co-author, you are supposed to know, and agree to, everything the paper contains. When a manuscript is revised, it is also assumed that all co-authors agreed to the suggested changes and the responses to the editor's, and reviewers' comments. As a co-author, this is not only your responsibility, but also your right (to know and agree to the above) — exercise it. Always ask for the complete copy of the manuscript before submission, and read it. Similarly, contribute to, obtain and read the revised versions and the response, and make your opinion known, especially if you disagree. If you do not do this, you have no one to blame but yourself. In a very grave case of disagreement, you can withdraw your authorship (before acceptance, naturally).

In a submission letter, if there is more than one author, it is to be clearly stated that the content of the paper is known and agreed in all respects by all the authors. No-one should then enter into any argument about apportioning merit or blame, should any question emerge about the paper's contents. In a submission letter, if there is more than one author, it is to be clearly stated that the content of the paper is known and agreed in all respects by all the authors.

Consequently, it is unwise to accept authorship, even if offered, on any paper to which one did not contribute. If you are a corresponding author, or the one who is organising the writing of a paper by a team, make sure that all the authors receive and read the manuscript before submission (or, at the very least, they declare that they have). In the case of any later dispute, you are then free of any eventual accusation. All this may sound paranoid, but very bitter stories have ensued because these aspects were neglected. Also, make sure that all the co-authors are fully informed about the development of, and eventual changes to, the manuscript.

The Corresponding Author's Role

In the case of a team co-authoring a manuscript, the editor always corresponds with only one of the authors. This author is called the "corresponding author". This is a service that can be performed by any of the authors, and is not linked to, nor attracts, any rank among the authors. It is not necessary, although it often happens, that the first author is also the corresponding author. There are many exceptions. The identity of the corresponding author should be agreed on before submission by the authoring team. This author provides a clerical service for the team and is the contact point for the editor and, upon publication, the outside world. However, the role is not merely clerical. The corresponding author often organises necessary revisions and, upon acceptance, submits the final copy. Page proofs are also sent to the corresponding author only.

Junior first authors sometimes shy away from this role. However, corresponding with editors and seeing the publication through to print is a skill that must be learned. It is a good idea to allow a young author to take this role, with senior, more experienced authors giving support as needed.

7

Listing the Addresses of Authors

The addresses of the authors are usually printed after their names, near the top of the article. This is to indicate where the work was done, and to provide an address for possible correspondence with the authors. This should always be a complete postal address.

Sometimes, by the time the publication appears, the author(s) have left the workplace where the published work was done. They must still write the address where the work was done give this address as their first address (to give credit to the institution where the work was carried out) but they should also provide a *current address*. The current address is where the author can now be reached.

The *address for correspondence* allows the editor to communicate with the author. This should be the address where the corresponding author can be reached. There can be multiple corresponding addresses, considering that the manuscript evaluation, review, etc. process may take several months. If the corresponding author is expecting to move during the probable assessment period (ca. 8 months from manuscript submission) and to have another address for at least one month, this address, and the period during which the author will be there, should be indicated. The corresponding address should also contain telephone and fax numbers, as well as an electronic (e-mail) address.

Routinely, journals now require that the corresponding author gives an electronic (e-mail) address and this is published with the article. With the increasing use of the Internet worldwide, contacting an author may be easier by e-mail than by other means of communication.

8

Significance of Keywords and Summary

The Abstract (sometimes called the Summary) is always printed near the start of the paper, usually immediately after the title, authors and addresses. This is, one can argue, the most significant part of a paper, because:

- this is the part that is read by most people, even by those who will, ultimately, not read the whole paper.
- most readers, including the first reader of your manuscript submission — the editor — will also start reading here. First impressions are important. Moreover, the editors know from experience that a bad abstract is rarely followed by a brilliant paper. Consequently, after reading the abstract, the editor will be close to forming a first opinion.
- an abstract is often reproduced by itself in various databases. This dictates that the requirement that the summary should be self-explanatory — it must be understood without reference to other parts of the paper. For many readers of the summary, the full paper will not even be available.

Occasionally, an abstract gains additional importance. Conference invitation, participation, and even financial support to attend, can depend on the abstract of a proposed contribution. In these cases, the conference organisers must make their decision based on the abstract only. So, a good abstract can influence organisers to offer conference acceptance and/or funds to support participation.

A good abstract is a mini-review of the paper. It states, briefly, the question/problem, the method(s) used, followed by brief results and the main conclusions. Some journals follow a system of numbered statements, or headings within the summary. A few provide detailed instructions (Box 7). However, even if the journal in question does not indicate the main aspects with headings, you have to follow the same structure.

Box 7. Nature's abstract-writing template

nature

How to construct a *Nature* summary paragraph

Annotated example taken from *Nature* **435**, 114-118 (5 May 2005).

*One or two sentences providing a **basic introduction** to the field, comprehensible to a scientist in any discipline.*

*Two to three sentences of **more detailed background**, comprehensible to scientists in related disciplines.*

*One sentence clearly stating the **general problem** being addressed by this particular study.*

*One sentence summarising the main result (with the words "**here we show**" or their equivalent).*

*Two or three sentences explaining what the **main result** reveals in direct comparison to what was thought to be the case previously, or how the main result adds to previous knowledge.*

*One or two sentences to put the results into a more **general context**.*

*Two or three sentences to provide a **broader perspective**, readily comprehensible to a scientist in any discipline, may be included in the first paragraph if the editor considers that the accessibility of the paper is significantly enhanced by their inclusion. Under these circumstances, the length of the paragraph can be up to 300 words. (The above example is 190 words without the final section, and 250 words with it).*

During cell division, mitotic spindles are assembled by microtubule-based motor proteins[1,2]. The bipolar organization of spindles is essential for proper segregation of chromosomes, and requires plus-end-directed homotetrameric motor proteins of the widely conserved kinesin-5 (BimC) family[3]. Hypotheses for bipolar spindle formation include the 'push–pull mitotic muscle' model, in which kinesin-5 and opposing motor proteins act between overlapping microtubules[2,4,5]. However, the precise roles of kinesin-5 during this process are unknown. Here we show that the vertebrate kinesin-5 Eg5 drives the sliding of microtubules depending on their relative orientation. We found in controlled *in vitro* assays that Eg5 has the remarkable capability of simultaneously moving at ~20 nm s^{-1} towards the plus-ends of each of the two microtubules it crosslinks. For anti-parallel microtubules, this results in relative sliding at ~40 nm s^{-1}, comparable to spindle pole separation rates *in vivo*[6]. Furthermore, we found that Eg5 can tether microtubule plus-ends, suggesting an additional microtubule-binding mode for Eg5. Our results demonstrate how members of the kinesin-5 family are likely to function in mitosis, pushing apart interpolar microtubules as well as recruiting microtubules into bundles that are subsequently polarized by relative sliding. We anticipate our assay to be a starting point for more sophisticated *in vitro* models of mitotic spindles. For example, the individual and combined action of multiple mitotic motors could be tested, including minus-end-directed motors opposing Eg5 motility. Furthermore, Eg5 inhibition is a major target of anti-cancer drug development, and a well-defined and quantitative assay for motor function will be relevant for such developments.

Nature's abstract-writing template

Because an abstract is often reproduced separately from the full paper, it has to be self-explanatory. Consequently, you should avoid using abbreviations, because they are understandable only by reference to the full article. Similarly, references to figures or tables are not allowed, because readers who only have access to the abstract cannot check or see the figure mentioned. For the same reason, references to published articles should be avoided. If unavoidable (for example, your paper is a direct reply to a published paper), a short version of the full reference should appear in the abstract. This short version should include the surname of the first author, the year of publication, the abbreviated name of the journal, the volume number, and the number of the first page only.

Style

Abstracts always have a word limit, usually 200-500 words. You cannot go beyond this limit, but it is not mandatory to use all of it. If you can clearly summarise your study in 150 words, you do not have to use 200. The important thing is that you should not list what was done, but concentrate on the results. The abstract centres on your own results, so it should be mostly written in the past tense.

The abstract is a summary of the paper, and there should be no statement or conclusion that is not in the paper. One should be careful not to include information that is not in the text (a surprisingly common error!). A good abstract is not a set of carefully cut-and-pasted sentences from the full paper; you must rephrase the same facts or statements that are present — usually in more detail — in the paper itself. At the end, the conclusions can be mentioned. These, however, should be meaningful. The statement that "The consequences are discussed" is neither very original, nor does it say much. This is the purpose of the discussion, after all. "More research is needed" is another meaningless conclusion. Avoid "throwaway sentences" such as these.

When to Write?

I suggest that the abstract is best written once the manuscript itself is finished. This is only a personal recommendation as I do this myself. If it helps you to structure the paper, you can start with drafting the main points but I find it unlikely that you can write an effective summary of your work before it is completed.

Research Highlights, Graphical Abstract

Several journals now require a tightly structured set of "research highlights". The relationship between these and the full summary is a little like that between the running title and the full title. Concentrate on your main results, and consider, carefully, how to shorten them because this section is limited by the number of spaces, i.e. letters and punctuation.

If required, a graphical abstract accompanies the research highlights. Both will appear on the website of the journal but not, usually, in the final paper. When facing such a task, think about creating a new figure rather than repeating one of your figures from the full paper. However, this is not a rule: in cases when your main results can be effectively presented on a graph, this graphical abstract can be identical to a figure that is also in the paper.

Keywords

Keywords serve to assist those who use various databases and search engines to find your paper. They are usually single words that mention some important concept or aspect of your study. The number of keywords is always limited, usually to 6-10 words (double words and, exceptionally, triple words are also allowed). These key words will be entered into databases and keyword lists.

It is a good idea not to use words that are already contained in the title, because effective title words are always used for the same purpose. Given the limited number of possible "pointers" to your paper, it would be a wasted opportunity to use the same word twice: once in the title, and again among the keywords. Several journals do not allow title words to also be keywords. An effective keyword is a word with a specific meaning or significance; words such as *study, change,* or *experiment* are not effective words in this context.

When considering keywords, imagine yourself as someone searching for your paper. What aspects are relevant? These can be locations, organism names, concepts, or method terms. Avoid fashionable, or too general, keywords — a reader searching for information on a certain topic will probably disregard the results of a search with hundreds of "hits". Thus, if you include a keyword that is frequently used, your article may be included, but in so large a result list that it does not help the searcher. Still, you can mention broader ideas, or concepts. If your title contains a species name, mention the family or higher taxonomic association. If there is only a scientific name in the title, include the common name. If it contains a location name, add the name of a wider region. Likewise, if you indicated a kind of habitat, add the more precise location as a keyword.

However, use this option in moderation — you do not have to use the maximum number of keywords allowed. You can write fewer keywords if you want, but you cannot include more.

Keywords are usually placed after the abstract — but check the journal instructions for precise placement of the keywords.

9

The Introduction and How to Write it

The Introduction is also an important part, because when a reader decides — after reading the title and the summary — that the paper is worth reading, most of them will start reading an introduction. It has to provide the neçessary context in order to make the new information understandable. If you fail to evoke the reader's interest, she will not read on.

Firstly, you have to clearly formulate the problem, the central question of the paper. You also have to provide some information about this principal question. What is known about it, what are the uncertain points? What theories have been suggested so far, what attempts have been made to answer the central question, and what is the available evidence? You also have to explain why this problem is important and worthy of study.

What is known about the studied phenomenon or problem? When answering this, you have to give context, and review the literature. However, you should avoid the temptation to over-emphasise the extent of your knowledge. You demonstrate good scholarship by *not* citing too many references (an increasing number of journals limit the number of references cited). The key word here is selectivity: you should cite literature carefully and sparingly. Only major papers need be cited, and a good reason to cite is because the discovery itself is relevant; the mere existence of the study. The reader is not so much interested in knowing who studied a given problem but, rather, in its relevance and outcome? Providing context is important to enrich understanding. A local, specific theme should be weaved into the "universal tapestry" of knowledge. How does this situation relate to other, similar examples in other studies, or parts of the world? Are there differences or similarities worth mentioning?

An introduction has a funnel structure: one should start with the more general aspects, and gradually narrow it down to the actual question, which inevitably will be a partial one. For example, a paper about the effect of urbanisation on ground beetles may well start out by briefly mentioning global change as a phenomenon, moving on to state that urbanisation is an element in global change. Among the multitude of effects, the impact on biodiversity of urbanisation is a rightful concern, but all diversity is difficult to study, so the study of invertebrate biodiversity is the focus. To estimate the effects on invertebrates in general, beetles can be used as indicators of effects on invertebrate biodiversity. Then you can move onto the specific hypotheses you tried to test — in your location, with your group, using your method (if relevant). This way, you link the general problem with your own study.

It is a good idea to formulate hypotheses. Introductions often argue that the phenomenon to be studied is interesting, and we do not know much about it — and it stops there. The author believes that this is sufficient justification. It rarely is. The "suck it and see" approach is not a justifiable one. There are so many unknown things of potential interest on our planet — why study this one? Try to be more daring when formulating hypotheses. Based on the knowledge so far, what is your expectation? What do you think will be found? Formulating the research hypotheses here will help you to structure the whole paper, and it is worth thinking about. Be careful with the hypotheses: if there are too many (more than five), your paper will lose focus. Also, be careful to make the distinction between your scientific hypothesis, and its formulation as a null hypothesis. Popperian scientific logic dictates that you should formulate a "null hypothesis", i.e. if the factor you suspect will be effective is, in fact, not applicable (nothing happens), because you can only disprove a hypothesis. Rejection is unequivocal; if you cannot disprove a hypothesis, it means that the theory does not clash with reality — so you can uphold your hypothesis. However, you can only invoke a new factor when the currently accepted factors no longer explain the phenomenon observed. Therefore, initially you must hypothesise that the factor suspected does *not* act on the phenomenon measured.

Writing about hypotheses, and making them explicit, may bring additional advantages. It may lead the study in new directions or make you aware of a missed approach or method and, thus, help to further develop the research.

In this section, you also have to explain the choice of experimental methods. Often several methods are available — why did you choose the one you did? Did you exclude others? Why? However, the choice of statistical methods is not explained here, but in the Material and Methods section.

During this gradual construction of your specific research questions, keep the expected readership in mind. This governs the detail you have to include — certain concepts are clear to specialists, while not to a more general readership.

In the last paragraph, summarise your results and main conclusions. You do not have to repeat the summary, or any other part, word-by-word. Avoid this. Paraphrase, but mention the main conclusions. You can refer to the hypotheses: which one was upheld, and which one was refuted? Some journals, especially more conservative ones, do not follow this structure, but I think it is useful for the reader to know the "point of arrival" in advance. As Ratnoff (1981) observed: "Reading a scientific article isn't the same as reading a detective story. We want to know from the start that the butler did it".

Style

The Introduction mostly deals with previously existing knowledge; therefore, the simple present is the appropriate tense. Sentence structure sometimes necessitates the use of past tense. You also have briefly to refer to your own results (they will be presented in detail in the results section), and, here also, use the simple past tense. Previous publications are mentioned in the simple present.

Beware of jargon. Jargon is a useful shorthand, but not to the (marginal) outsider. Consider the job advertisement from a US journal (cited in Day, 2006): "NIH is an equal opportunity employer, M & F". Day asks the rhetorical question: does M & F stand for muscular & fit, hermaphroditic, musical and flatulent, or "mature and in his fifties"? In fact, neither: it refers, in jargon, to the relevant piece of US legislation, the non-discrimination of applicants based on sex, age, religion, sexual orientation, etc. Remember; jargon is for the "insiders" — but you have to inform the novice and the outsider as well as the expert.

When to Write?

The Introduction is usually one of the first sections to tackle. It is also good to formulate this section (even if not in its final form) while the investigation is still in progress. The whole team is still together, which makes the life of the corresponding author more painless. The co-authors are committed to the paper and will assist more willingly during the work, rather than 6 months after the completion of the project. It is also likely that you had to prepare a research application, or research plan, before you could start on the research. Thus, information about context, some literature, and other elements of introduction may be at hand — use these details at this stage.

10

Materials and Methods Section: Its Aim and How to Write it

Although traditionally, this section is only called "Material and Methods" (rarely: Study Site, Material and Methods), it can be composed of the following parts: study site, study organism, material, methods, statistical evaluation.

The aim of this section in scientific papers is to enable readers to assess the reliability of your work, and to be able to repeat it for verification if they want to do so. Science is about unearthing nature's laws, and the cornerstone of the scientific method requires that experiments are repeatable: if the experiment is repeated under the same conditions, the same result should be obtained. A material and methods section should give enough detail to evaluate and, if needed, to repeat the experiments reported in the article.

You should carefully consider your potential readership. This allows you to provide enough, but not superfluous, information. Once you have reflected on what can be assumed as known by this readership about your setting, organisms, methods, etc., you can give detail accordingly: not too little, and not too much.

During peer review, this section is closely scrutinised. If the reviewer is in any doubt that the experiments are repeatable, or that the methods are appropriate, the manuscript will be rejected as unreliable, no matter how wonderful the findings are.

Study Site

When describing your study site, consider your potential readership and give details accordingly (geographical particulars, history of the site, location, co-ordinates, maps). The aim is not to enable the reader to find your sampling plot, but to give a general understanding, a "feel" for the environment you worked in. Information on habitat, with photos, maps, drawings, is often useful, or wholly necessary.

Study Organism

Here, you should name all the species, strains, cultivars or races that were used in the experiments. You should also give precise information on their origin, storage or husbandry, including temperatures, photoperiod, feeding regimes, control, etc. Depending on the readership, you should consider giving other background information on life history, and the organism's distribution in nature. If there is a long list of organisms or strains, consider preparing a table with this information.

Materials

Here, you should list all the materials necessary for your experiments. Give exact names, not generic or trade names, of chemicals used. Give a source (manufacturer with location) if the chemical in question is delicate (e.g. an enzyme), or rare, or its quality is critical. This would give additional information to the reader. This is, however, neither advertisement nor endorsement (for legal reasons, this should often be made explicit in the paper — see, e.g. the US public organisation policy: disclaimer: "The mention of any trade name does not constitute endorsement by XXX organisation"). For equipment used, give the name, specification/type, manufacturer, and conditions of use.

Sampling Methods and Measurements

Here, you should detail the procedures: how did you perform the observations, measurements, experiments? How many times, under what conditions? If you use a new method, give all the details necessary so that the reader can repeat your experiment from reading this section. If you used a published method, a reference to the original publication, preferably the one that first published the method, is usually sufficient with minimum description. If you modified a published method, detail the modification only. If the method is published, you should cite it — but consider *where* it was published? Is it a frequently used method? When was it published? A rarely-used method, published long ago in an obscure journal, needs a more detailed description than a much-used, current one. If the original publication is not widely available, you will have to provide detailed description. Editors often welcome more detail, especially if the published method is not in very wide use (with the appropriate reference, naturally). If you modified a published method that is widely available, detail the modification only.

When describing the procedure, be aware that only SI (Système International) units of measurement are allowed. A few units in common use are not official SI measurements and they cannot be used. Also, be aware of the precise use of measurement units — for example, in common use, weight is often given as grams, kilograms, etc., but these are units of mass, not of weight.

Any larger set of samples, measurements, or experiments will have the occasional error, a missing sample, a lost or mislaid tube. Do not keep silent about them. Indicate, clearly, how you dealt with errors, missing data, missing traps. This will not decrease your credibility — on the contrary.

Evaluation Methods/Statistics

Data will mostly be evaluated by using a statistical program. In most cases, a reference to the program (indicate the version used) is sufficient; give detail only if the method used is new. However, avoid the neophyte description: what's new for you may not be new for readers. An experienced colleague can give advice on this matter. In general, it is always a good idea to discuss your chosen statistical method with others. Here, you should give a reason for the choice of statistical test, as well as stating how you tested the eventual conditions for using the chosen test (testing for assumptions for a given statistical test). The mention of the use of a commercial statistical program naturally assumes that you have valid access to the program in question. It is not unheard of program developers to search for the mention of their product in the literature to find out about illegal use.

Style

Be careful with details when writing a material and methods section — your reputation is on the line! The reader was not by your side when the studies were done, so she will use the detail and clarity of this section as an indirect indication of your reliability and thoroughness.

A common error in this section is not offering enough detail. This does not happen because of the authors' desire to hide anything — it is simply a mark of routine: many parts of the experimental protocol may become almost routine, and the small details are forgotten as they never change and are taken for granted. When the description is prepared, these details, vital for others, are often not included. A good test is whether a colleague, on reading the section, thinks she can repeat the experiment based on the given description of methods. Such a check is useful, because the writer often is too close to the methods, having done them countless times during the experimental process and, thus, omits some obvious but important, detail.

Specifically, take care with numbers, spelling, and punctuation. In this section, many "strange" names will occur: of chemicals, organisms, strains; concentrations, times and units of measurement are important. Meticulousness is the key word here: if you cannot be trusted to do simple things well, such as describing a method that you used hundreds of times, can you expect the readers to trust you when it comes to more significant and complicated aspects of reporting your research?

The order of description should be chronological; the description of what was done first should precede the later actions. However, you have to first mention all study sites, then all organisms, followed by a full list of all materials used, experiment-by-experiment and so on. Thus, if someone is only interested in all the details of, for example, your second experiment, she will have to jump from one part of this section to another. This seems a small price to pay for a consistent structure, which is followed by most journals.

This section describes your own work and, thus, the past tense is used, mostly, in this section. When describing the details, beware of the syntax. The following description is taken from Day and Gastel's book (Day and Gastel, 2006), who, tongue-in-cheek, called it "the painful method": "After standing in hot water for an hour, the flasks were examined". I hope this was not performed as the sentence implies — probably the flasks, and not the researchers, were standing in hot water that long.

When to Write this Section?

It is best to start writing this section first, possibly even while working on the experiments. Otherwise, many details will be lost. Details and precision are vital here, and they are much easier to document during the work, or soon after, than weeks or months later. Additionally, there is often a practical reason, too. Most scientific work is done in teams; it is much easier to convince the team members to write their respective methods section while they are doing the work, or soon afterwards. Once the experiments are completed, and the team moves on to further projects, writing a complete methods section will take longer, and be done less satisfactorily.

Meticulousness pays, because, as stated above, reviewers are often of the opinion that if you cannot be trusted in doing simple things, you cannot expect trust in significant and complicated aspects of research. Science, in the view of many of its eminent practitioners is, after all, "99% perspiration and 1% inspiration", so precise work, and the ability to describe things accurately, is a necessary condition of credibility. Science may well comprise a lot of precise work and fewer grand ideas; you prove your mastery of the methods applied by being able to describe them with clarity, in sufficient detail.

11

Writing the Results of a Scientific Study

This section is a key part of any scientific article; indeed, it is, the very reason of for writing the paper: the presentation of significant, new facts. Because of this, everything hinges on your results. If you do not have good enough results, convincingly presented, no matter how well the other parts of the paper are written, the manuscript will not be accepted for publication by any reputable scientific journal. The results must be new, possibly significant, compellingly represented, and the claims well-supported by evidence. This is the part where the new information appears, and one has to be very careful in about how to present this new knowledge.

The results are not a simple presentation of the outcome of your measurements. They have to be organised and interpreted to ease the task of the readers, so that they can most easily understand the novelty and the nature of the new information. So, a perfect Results section does *not* read: "The results can be seen in Figure 1". Results — other than occasionally — do not speak for themselves. You have to organise the results into a sequence, possibly so that it is consistent with the problem statement, the starting hypotheses, and so on, in turn. You have to present:

a) the big picture, an overall description of the experiments. What did you manage to prove?

b) results that support the claim you make. How do they corroborate the claim you make under a)?

When writing, detail significant data, not insignificant ones. If a variable was ineffective, mention it, but do not give excessive detail. Not all data need a table. Also, be aware of the saying: "the absence of evidence is not evidence of absence"; detail what did you not find as well.

Here are a few things you should avoid.

- Do not start with a forgotten method — include that in the appropriate place in the previous section.
- Do not present material as results. Obtaining your study material is necessary for your research, but this is not a result. It does not belong to the results section; for example, how many birds you captured for your study is not appropriate, only the outcome of the actual measurements. If you had to identify the material collected, then it becomes a result.
- Do not try to present all your data. The fact that you measured more data than anyone else will not increase your reputation. In effect, the opposite may be true: "the compulsion [...] not to leave anything out does not prove unlimited information; it proves lack of discrimination" (Day, 1998).
- There should be no double presentation of the data. Do not repeat your data in text *and* tables/figures. This does not mean that occasional values, presented in a figure or table, can never be mentioned. Precise values cannot be read from a figure, for example, so if a value, also present as a data point on a figure, is important, it can be mentioned in the text. In general, the most economical method should be used to present your data. For further discussion on how to decide on the most suitable form for presenting data, (text vs. figure vs. table), see Chapter 15 and Chapter 17.
- Do not leave the reader to find the meaning and analyse your data on her own. The reader will see your new data for the first time, and needs your help to quickly grasp their nature, meaning, and novelty. You are the best guide for her, and do not shy away from this responsibility.

When describing the results, follow the "from macro to micro" principle: first make a general claim or point, then illustrate or prove this by giving more detail. Present details, statistics, etc. that support your argument. Point to significant trends and facts among the numbers. This way you can direct reader's attention. There is no need to interpret the results — that should be done in the following section, the Discussion.

Do not yield to the temptation to combine results and discussion. Authors sometimes resort to this with a claim that "it is easier to understand the flow of the argument". Interestingly, however, most such authors do not seem to have a clear line of argument to present. On the contrary: the argument is often muddled, which is not made clearer by immediately mixing it with interpretation. Often, it is not easy to separate the authors' own work from that of others. However, it is very important, to make it especially clear what they measured or observed themselves, i.e. what is the empirical base, and what is the speculation or interpretation. This is extremely important for colleagues with a smaller publication record — it is to their advantage that the world can clearly see what they did (results) vs. how they argue (discussion).

The sequence of the presentation should also be consistent in the sense that, if there are several experiments carried out, the results should be presented in the same sequence as they were described in the previous sections, the Introduction and the Material and Methods.

Results contain new information, new facts. Every statement should be supported by facts: a figure, a table, a number, or a statistic. Most of the figures and tables contain results, and are to be placed in this section. All figures and tables should be integrated into the narrative. Do not simply claim that the results of one experiment can be seen in one or more figures/tables. For understanding, do not rely on the legends, either, even if figures and tables should be understandable without reference to the text. Link text statements and their evidence into one narrative.

Presenting Statistics

When a statement is supported by a statistic, the conventions are precise and rigid: give the name of the test, give the test statistic value, followed by the degrees of freedom, and the level of significance. For example: "XZ was significantly larger than FF (Student's t-test, $t = 5.43$, d.f. = 114, $p = 0.00014$)". In general, if your statistical significance was calculated using a program, the output should be a precise value of the probability of error, p. The notation $p<0.05$ indicates that the evaluation used a statistical table, where only levels can be assessed (i.e. $p<$ *or* $>$ 0.1, 0.05, 0.01, 0.001, etc.). Beware of false precision: every measurement has a sensitivity, and presenting the mean of measurements to five decimals implies a very high sensitivity (0.00001). Do not present more decimals than the sensitivity of your original measurement. Some allow that the number of decimals in means can exceed that of the original sensitivity by one: if the original measurements have a sensitivity of 0.1, the mean and the chosen measure of variability (s.d., s.e., confidence interval, etc.) can be given as x = xx.01.

Box 8. Presenting numbers

- Numbers <10 are usually written out in letters; if they are >10, Arabic numbers are used
- When the numbers represent measurements (they have a measurement unit), always write them as numbers, and never in words
- Sentence does not start with a number in Arabic. You cannot write "200 birds were captured...". If it is necessary to start a sentence with a number, always write it out in letters, irrespective of its magnitude: "Two hundred birds were captured..." or "One unicorn was captured...."
- Observe the writing of the number and its unit — sometimes there is, sometimes there is no space required between a number and its unit.
- Numbers smaller than 1 should always start with a zero, i.e. 0.123 and not ".123"
- When reporting a range of values, use the "to" rather than the mathematical symbol. The latter can be misinterpreted as the negative sign. Write "1988 to 1996" or "-23°C to 18°C". When writing page numbers in the reference section, though, use the "en-dash" symbol: Ecology, 24, 133--145.
- Use the official Système Internationale (SI) units only
- Be careful about mixing up scientific and everyday use of properties measured (e.g. weight vs. mass)
- Beware of exactness! Measurements always have a level of precision (to be indicated in the material and methods section). When one deals with measurements, the mathematical absurdity "$1 \neq 1.0 \neq 1.00$" is true. The last number (1.00) indicates a measurement sensitivity which is a hundred-fold higher than the first one (1).
- When presenting means and other numbers calculated from the original data, retain the sensitivity of the original measurement. If your measurement exactness was 0.1, do not give its mean as 0.13333. If your

sample size is n <50, give only whole numbers, i.e. 15%. When your sample size is n <20, give the actual numbers, and no percentages.

- Singular-plural may have different meaning:
  10 g was added — as a single dose/quantity
  10 g were added — a total of 10 g was added, but in several doses

Also, statistics must be meaningful, which is not the case in the following statement (from Day, 1998): "in 33.3% of mice, the treatment was effective; no change was observed in the condition of the other 33.3%; the third mouse escaped".

Grammar and Style

The Results section usually contains no references, because it describes your own, new work. Use the simple past tense. Here, your own results are being presented for the first time, and convention requires that they are not yet treated as established facts, and the use of the present tense is not appropriate. This section will certainly include numbers, for which special rules apply (Box 8). These may vary by journal — check the necessary format and follow the requirements.

Concerning style, there are overriding necessities: crystal clarity and simplicity. Precise language and clear, simple statements are not only highly appreciated stylistic values of the English language — precision and clarity make the understanding of new information easier. A common mistake is to write overlong, convoluted sentences that, also, sometimes contain infrequently used, foreign terms. Neither is optimal, nor is it necessary to the degree used. Ideally, one sentence should make one point or statement. Only occasionally can an interpretative clause be added. The simplicity of the coding principle is again invoked: this is the part that is new, and contains information that is known to no-one but you — at least before the article is published. The world at large will have an easier task to understand the magnificence of your new results if they do not also have to struggle with a convoluted style. Simple sentences indicate profundity of thought.

12

Writing the Discussion: Things to Remember

This section is hard to define, and it is also hard to write. Often, the Discussion section is the most verbose of the manuscript, and can pull down the whole paper. Most rejections happen due to a faulty discussion. Consequently, this section becomes the most important one for the aspiring author.

Authors often follow the "squid technique" (Day, 1998): being unsure what their data mean, they try to hide in a protective cloud (of ink!), confusing the issues that are pertinent to the results.

A discussion should provide the answer to one simple question: what do the data mean? When attempting to answer this question, there are several "dos" and "don'ts". First, what to avoid in a discussion:

- Do not repeat results. This does not mean that the overall conclusions cannot be mentioned, but reciting the results in different words is unnecessary and superfluous.
- Do not introduce new results. In some cases, the relevance of certain results becomes evident only during the writing of the discussion. Make sure that these are mentioned in the results section and, also, that their relevant methodological details appear in the Material and Methods section.
- Do not pretend to have solved everything. The significance of the findings should be mentioned, and as authors, we genuinely believe our results will shed light on new areas. It is unlikely, though, that all relevant problems have been solved, and it shows no wisdom to make such claims.
- Do not finish with throwaway sentences. Towards the end, the overall significance of your findings could be mentioned. This, however, should not be something very trivial, such as "more studies are needed". Mentioning

"significant new avenues for future research", or "indicating significant human influence on XX ecosystem" are also of the "throwaway" kind: they have been made so often that they no longer mean very much. It is best to avoid them.

- Do not try to discuss every possibility, especially if speculative. Your results may have connections to several other areas, facts, and theories. Be modest when trying to discuss these.

However, do:

- Present principles, relationships, and generalisations that follow from your results.
- Refer back to your central problem and, also, write about what was *not* found, corroborated, or supported from your predictions. Point out any gaps or inconsistencies that continue to exist, or that your results indicate.
- Show how your results agree and differ with previous work. Both aspects are important for the further development of the field. Additional support of existing facts, theories, or ideas with a new experimental setup or system is important. Also, do not be silent if your results are different from previous findings. Try to identify and discuss the possible reasons for this. Be cautious and precise when invoking potential factors.
- Explain the significance of your results. You are the best person to understand, and explain, the significance of your findings. Avoid the "so what?" response of potential readers. They may not be willing, or may not be in a position, to go further to analyse your results, and find out about their significance. You know your data well, and you probably have a good perception of their importance. Be honest, and modest — but do not shy away from pointing out the importance of your results.
- Discuss the theoretical and practical implications. A piece of practical work often has theoretical implications, and vice versa: a theoretical work may suggest practical

applications. Point these out; try to think of the possible other uses of your work.

- Present the "new picture". Again, you are in the best position to appraise whether the new results you present in the results section lead to a new situation. It is your privilege, but also your duty to write this, if pertinent.
- Summarise evidence for each conclusion. Do not leave this to the reader, even if you believe it to be obvious. Do not assume anything, and do not leave this to the imagination or intellect of the reader.

A common error is described by Day (1998) as "a clear stream of discussion ends in a swampy delta". Finish the paper with a clear statement. There is no need for a cosmic conclusion, but a well-formed, precise statement ends the paper nicely.

Style

The style of discussion is more complex. The tense of this section will switch between present (reference to published knowledge, with a citation) and simple past (when mentioning your own results). Occasionally, other tenses may occur in this section.

A discussion should be kept in proportion with the results. Generally, a discussion longer than twice the length of the Results section often risks "discussing the findings to death". Write using your own words; do not repeat the wording from earlier sections. There is no need for far-reaching conclusions; you will be able to illuminate one area. Your conclusions can be buttressed by your facts in that one area — but if you extrapolate to a bigger area than your data allow, you may appear foolish to the degree that even your data will be doubted. "Display your small piece of truth — leave the whole truth to ignoramuses, who proclaim its discovery every day". (Day, 1998).

The simplest statements evoke the most wisdom. Fancy language and technical terms may be used to disguise shallow thinking. Try to write simply.

13

Writing the Acknowledgments Section and Appendices

No one is too big to be courteous, but some are too little.
Ralph Waldo Emerson

Scientific research, like any other human activity, relies heavily on co-operation. An important part of this is the assistance we obtain from, or provide to, colleagues. This can take many forms: loan of equipment, help to learn a method, discussions of experimental plans, statistical advice, support in the field, and so on.

Often there is no payment, nor any reward, to the people who provide such help. Co-operation is commonly and courteously acknowledged by thanking the people who helped. The same rules apply: if in everyday life, you would say "thank you", do this in your paper, too. The appropriate place in a scientific paper for such gratitude is the acknowledgements section. The Acknowledgements appear after the main text (Discussion, or Conclusions) and before the Reference List.

Who should appear on this list? People who helped you in significant ways to complete the work reported in the article. This includes technicians, field assistants, authorities who gave permission or provided access to resources, colleagues who commented on the manuscripts, or helped you in significant ways. However, this is not really the place to thank your partner, or the coffee lady in the department, even though you feel you could not have completed the work without their support/ coffee. Buy them a bouquet of flowers or a bottle of wine instead, and say thanks in person.

Be aware that this is not a "surprise present", so the person to be thanked should know about it, and agree to it in advance. Show the wording, too. There are good reasons for this. The person may feel your acknowledgement is too much — or too little. Scientists often co-operate "on the ground", but their superiors are not always happy about this. Maybe the person concerned would have preferred to remain anonymous, or maybe she thinks she deserves to be a co-author.

The Acknowledgements section is also the place to record sources of financial support. Most funding agencies require that you mention them in any publication that emerges from work done with their support. Here, you should also note that you complied, if necessary, with regulations, obtained necessary permissions, and so on. If the work forms part of your MSc or PhD thesis, this should also be indicated here.

Many workplaces, field stations, and programs have a running, numbered list of publications. All publications obtain a number, and these forums request that you list this number on your paper. Normally, it appears in your acknowledgements (the other, less desirable, option is in a footnote on the first page of the paper).

Authorship decision principles can be (and sometimes are) mentioned in this section (Tscharntke et al., 2007). If there is no section devoted to detailed author contributions, they should also be listed here.

Before publication, several people will probably read and comment on the manuscript. It is also courteous to acknowledge them. It does not matter whether you accepted their suggestions (you are not obliged to accept any of them) — you should thank them for their time spent reading your work. There is no need to qualify your gratitude, nor the advice. It would be silly to "thank XX for her brilliant suggestions". It goes without saying that whatever any reviewer's opinion or advice, and irrespective of your acceptance or rejection of them, the responsibility for the content of the paper is solely that of the authors'.

As for the style, be as brief as possible, to the point and direct. One should not "wish to thank" but simply "thank" someone.

Appendices and Extra Detail

Whatever does not belong strictly to the flow of argument in the paper, but is still important to the paper as a whole, can be presented as an appendix, or a series of appendices. This can include the description of complicated procedures, listing of programs, more detailed descriptions of models used, and long lists of large bodies of data. Several journals have established a freely accessible data repository, or data archive. These are linked to the website of the journal. Journals encourage their use and this will probably become more widespread in the future. Researchers might also make their data openly available separately from the journal (for example, in an institutional repository). Printed appendices are destined for extinction.

In any case, if you want to include appendices to your paper, be prepared to have to defend their inclusion. The desire to include an appendix is not certain to result in a fight with the editor, but have a good justification as to what purpose the appendix serves, and why you want it to be included. Editors are always on the lookout to shorten papers.

14

Citing the References

A manuscript containing innumerable references is a mark of uncertainty rather than a mark of scholarship
Day and Gastel, 2006

Known facts have to be mentioned in a primary paper, to help the readers to understand the new information. Known facts are marked by tense (present) and by a citation: indicating the discoverer by citing the work where the discovery was made public. Details of the cited works should be listed at the end of the paper, so that interested readers can find more information on the aspects cited.

Reference lists cannot, and should not, be exhaustive. This, as expressed very succinctly in the quote above, requires that you exercise your judgement about importance. Cite only significant, citable references (an item that has a digital object identifier (DOI) can be cited).

Also, be aware that different types of publications carry different weight. The primary paper is considered the most reliable source of information, because it presents the empirical evidence for the discovery or statement. Review papers and books closely follow — but they are not primary publications, and they rarely present the empirical evidence in full detail. At the other end of the spectrum are the various non-primary publications, yearbooks, websites, and personal communication. Choose carefully among them.

Do not cite a publication that you have not seen. This is a dangerous, if not rare, practice. Judging from "propagating errors" (checking for the reappearance of printing errors in subsequent citations of the same paper), Simkin and Roychowdhuri (2003) estimated that only ca. 20% of the authors citing a famous paper had actually read the original publication. This is dangerous because, when citing a paper, one must summarise the main points, or the reason for citation. So, the content of the whole paper will be summarised in a few words (by someone else other than the author); hence, this is an interpretation of the original findings. A subsequent citation, if it is based on this short, interpretative summary, will rephrase this summary, and distortion is inevitable. As credibility and precision is very important in science, the consequences can be very unwelcome.

If citing a paper that you have not seen is unavoidable, the format should be: XYZ 1874, cit. BB 1999. In the Reference List, give bibliographic details of the citing publication — the one that you have seen.

References should, preferably, be of sources that are obtainable, so that any reference could be checked if one so wishes. Therefore, it is wise generally to avoid citing unpublished data (unobtainable), unpublished manuscripts (unobtainable), abstracts (no proofs are presented to substantiate any claims), theses, government reports (both are published in a few copies only, and frequently difficult to obtain — although they are increasingly made available electronically), personal communication (person rarely available to testify). If you want to cite someone's opinion that has been given to you directly, such personal communication should be cited giving the full name and workplace of the person supplying the information, followed by the words "personal communication". This should not be included in the reference list. If you cite a personal communication, it is wise to keep a printed copy of that correspondence for your files.

Papers that are "in press" can be cited. This, however, does not equal, submitted" or in preparation". Many journals require that any "in press" citation is accompanied by a letter from the editor of the publishing journal clearly stating that the paper cited has been accepted for publication in that journal. This equally holds for your own, or for others", "in press" papers.

Citation Conventions

In-text citation:

In the text one should usually give the surname followed by the year of publication in parentheses; otherwise both the author's surname and year should appear after mentioning the relevant fact from the paper cited.

In the case of one or two authors, the surname(s) are always given in full: Smith (1980), Smith and Jones (1981). When there are more than two authors, the rules may vary. Some journals require the listing of three or four authors at first mention. Subsequent mention of the same paper is by the "first author et al." system: Smith, Jones and Little (1982) at first mention, Smith et al., (1982) later. More than four authors are usually cited as "Smith et al. (year)" even on first mention. If there are multiple citations in one sentence, the sequence is normally chronological, but sometimes alphabetical. You should check the journal requirements.

Do not place all citations at the end of the sentence. This can be difficult to read in the case of a composite sentence. Citations at the end of the sentence mean that all cited items state all the things mentioned in the citation. Often this is not the case: one paper is cited for the first part of a sentence, and another one for the latter part. The citation that supports one part of a composite sentence should be clearly indicated by placing the citation immediately after mentioning the fact.

Style

When citing a paper, follow an effective citation practice. This means that the relevant paper has measured, evaluated or proven the fact for which you cite it. Do not cite a paper simply because the desired sentence is somewhere in that paper. The reason for citation should, in most cases, be because of the results section of the cited paper.

Avoid the evaluative citation style. You are not asked to assess others' intellectual capacities. Do not write sentences like: "Jones' very elegant paper (1998)", or "Smith's long-discredited theory (1966)". If you cite a paper, do not judge it, but give a clear reason why you have cited it.

The Reference List

At the end of the paper, you should collect and present all bibliographic details of the cited publications so that readers who want to find further information can find the sources of your citations. Every item that was cited in the text must be listed here with the required details, and everything that is on the Reference List must be cited in the paper at least once (you can cite the same article several times).

While there is little difference of opinion over what bibliographic details are necessary to find a literature item, the formats in which one should present this information are bewilderingly varied and, sometimes, illogical. We scientists have, so far, failed to come up with a uniform way of presenting bibliographic information. I hope for, but do not expect, a future when there is a unification of citation format.

The three most common are the following, based on the CBE Manual (see Box 9 for examples):

The "uniform requirements for biomedical journals" (Vancouver system).

This system lists the cited items in the sequence of citation in the text, without considering the author's name or year of publication. The citation itself is an Arabic number, usually in superscript. This system is followed by the *Nature* group of journals, *Science*, and several other reputable journals. Despite its name, however, it is far from "uniform" — many journals follow other formats.

The Chicago Manual of Style — alphabetical.

This forum recommends the alphabetical listing of cited items. According to this system, the cited publications are first ordered by the alphabetical order of the first author's surname. After this, a chronological order is used. In case of overlap, the second author's surname is considered, and so on. In the case of an identical author team and publication year, letters indicate the difference: "Magura et al., 2010a; 2010b".

The Council of Biology Editors Manual — the alphanumerical listing of publications.

This style differs little from the previous one, except that, after arranging the publications alphabetically, they are numbered. In the text, only these numbers are included, and they point, unequivocally, to the relevant citation. Journals following this system argue that this improves readability, because the text is not broken by frequent parentheses and authors names and years, which are, from the point of understanding the argument, irrelevant.

Box 9. Samples of citation styles

Paper:

Ernise DJ, Kluge AG 1993. Taxonomic congruence versus total evidence, and amniote phylogeny inferred from fossils, molecules, and morphology. Molecular Biology and Evolution 10, 1170-1195.

J. S. Carr, A. T. Tokunaga, J. Najita, Astrophys. J. 603, 213 (2004)

Book:

Dressler RL. 1981. The orchids: natural history and classification. Harvard University Press, Cambridge, MA, USA 252 pp.

Book chapter:

Danchin E. 2001. Public information and breeding habitat selection. In: Clobert J, Nichols JD (editors), Dispersal. Oxford University Press, Oxford, UK. Pp. 243-258.

Website:

van Frankenhuyzen K, Nystrom C. 2002. The *Bacillus thuringiensis* toxin specificity database.

For journal articles, a complete citation includes the surnames of all authors, plus initials, the year of publication, the title, the name of the journal, the volume, and the numbers of the first and last pages of the cited article. Today, due to seemingly ever-larger teams and, thus, ever-increasing number of authors on papers, there is a limit to the number of authors to be listed, which you should check in the "instructions to authors" of the journal. The issue number is not usually needed, because volume pages are continuously numbered. For the precise format, check the journal requirements — you must follow them.

For chapters in compilations, collections, or books, the above details should be given but, also, the names of the editors, the title of the whole volume, and the publisher details. For an Internet resource, try to find all the above plus the full address (the URL), and the date of access. This is necessary because Internet-based information is ephemeral; the half-life of such material varies by discipline but can be as low as 1.4 years (Oguz and Koehler, 2016). If available, use the DOI of the document — this makes it more easily traceable.

There used to be abbreviation rules for journal names, and some journals still follow them. If in doubt, write out the full name, and the editor will help you to find the appropriate abbreviation. Follow them closely, because if your reference style is very different from the required format, your manuscript will be returned for re-formatting before it is even considered for review.

An estimated 60% of published papers are not published where they were first submitted, and it is a rarity that journals follow an identical reference format; authors therefore often have to re-format their manuscript before sending it to another journal. Consequently, it is wise to have a full database, with all possible elements present for a citation, because it is much faster to delete superfluous things from a manuscript than to type in missing ones. For your database, collect all bibliographic detail.

Reformatting a manuscript is necessary, but not creative, work; reformatting a reference list is a thankless and, potentially, unnecessary task. Fortunately, all major word-processing programs can link up with one or more literature databases, and can import citations from there. Such literature databases include, for example, Reference Manager, EndNote, or Zotero. Obtain and learn the use of one of them — they are more or less equivalent. They can be linked up to large Internet-based databases, such as Web of Knowledge or Scopus, and bibliographic data of selected articles can be downloaded directly.

The real advantage comes when you want to include citations. You have to open both programs, and can import the full citation, in a pre-defined style, into your manuscript. While you have to check the precision of these imported citations, they ease, tremendously, the compilation of the reference list, and save a lot of hassle and time if it must be reformatted for another journal.

15

Graphs and Figures

Clear graphics aid, and show, clear thinking about what data mean.
Valiela (2009)

A good graph is a well-designed presentation of interesting data, communicating complex ideas with clarity, precision, and efficiency. It gives the viewer the greatest number of ideas in the shortest time, with the least ink in the smallest space. Statistical graphics can: show the data, and induce the viewer to think about the substance, and not about methodology, design, or technology (Tufte, 1994). They also avoid distorting what the data have to say, present many numbers in a small space, and make large data sets coherent by encouraging the eye to compare different pieces of data, while revealing data at several levels of detail. A good figure also serves a clear purpose: description, exploration, or tabulation. Figures also have to be closely integrated with statistical and verbal description of a data set.

Terminology

Scientific graphs can take many forms. The most common ones depict the relationship between a response variable and one (exceptionally more, see later) explanatory variable(s). In their most common form, they are two-dimensional, using the Descartes coordinate system. Thus, there is usually a horizontal axis (often denoted the "x axis", Figure 1), and a vertical (or "y") one. According to tradition, the independent (explanatory) variable is presented along the horizontal, and the dependent (response) variable along the vertical axis. The axes are usually drawn to correspond to the range of values along them. The area thus indicated is called the *data rectangle*. Axes have *ticks* to indicate the scale, and *tick labels* to identify the scale values. The axes themselves have to be named using *axis labels*. The data series are represented by various symbols, and their meaning is given in the *legend* or *key*. The figure is accompanied by a *caption*: text that is printed below the figure, describing what is pictured. There can be multiple *panels* on the figure, making it a *multi-panel figure* (see below).

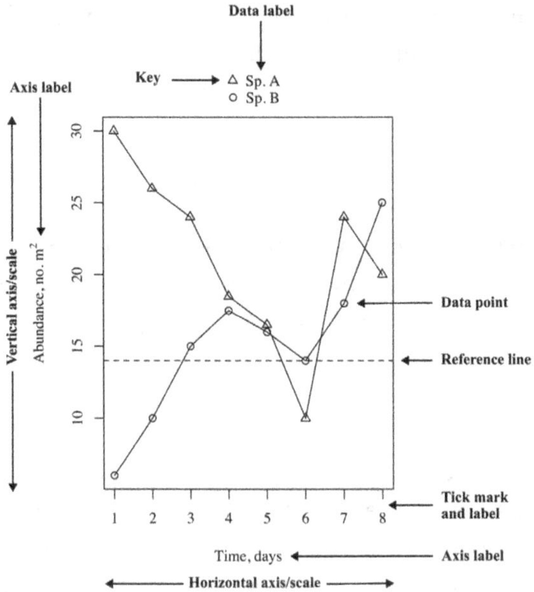

Fig. 1 Figure terminology. The square area defined by the two axes and their parallel lines is called the 'data rectangle'. Image by author (2020).

General Rules of Scientific Graphs

There are general design rules for figures that are worth mentioning here. The first is the *prohibition of double data presentation*. A set of data can be presented in only one way — either in text, on a figure, or in a table. Single values and trends can be mentioned and discussed in the text, but larger parts of the whole dataset cannot be presented in more than one way.

A second rule is that figures, together with their captions, have to be *self-explanatory*: the reader should understand *what is pictured* on the figure, without reference to the text or to other figures. Note that the interpretation of the figure does not need to be given here — that goes into the text (of the Results section, see Chapter 11). All axes have to be identified, including the precise naming of the measurement units, even if this may seem redundant (such as "time, days"). The data rectangle is for the data only: viewers instinctively try to interpret everything within the data rectangle as coded data.

The *simplicity of coding* principle has to be mentioned here, too: figures are coded information *par excellence* and, if there is a new symbol, code, or colour, which constitutes *visual novelty*, this should be understood first, before the data can be interpreted. This gives rise to rule no. 3: *keep visual novelty to the necessary minimum.*

Principles

There are three basic principles in the graphical presentation of data: economy, integrity and clarity. Figures should be designed to present the data in a clear, uncluttered, and honest way.

Economy

Presentation of data should be as simple and clutter-free as possible. However, this should be realised with economy: a figure should neither be bigger nor use more ink than necessary. The principle of maximising data *and* minimising ink (maximising the data:ink ratio) was suggested by Edward Tufte, and was aptly named the "Tufte Principle" by Valiela (2009). The principle of economy should also be considered when deciding the best way to present your data: in text, in a figure or in a table? The option occupying the least space is usually preferred.

An important first point here is that uninformative, decorative motives should not be used. Unnecessary decoration, shading, cute pictures and other uninformative elements justly acquired the name "chartjunk" (Tufte, 2003). They mostly serve to disguise shallow thinking, and they betray a lack of belief in the data and a profound disdain for the intellect of the reader. Unfortunately, most graphical computer programs offer a vast range of chartjunk; ignore those.

Figures are expensive, and they should convey complex, often multi-layered data that need, and reward, scrutiny. Therefore, space is at a premium. Data should stand out, and everything on a figure should serve the data.

In the first instance, this means that the data rectangle should be filled with the data, to the largest extent possible. This requires a careful selection of the intervals to be included on both axes. They should only encompass the range within which there are actual data, and not beyond. Doing otherwise would simply be a waste of space. Using semi-logarithmic or double logarithmic scale often allows a more even fit of the data within the data rectangle (Fig. 2).

The most important additional information consists of the tick marks and labels on the axes. They should be sufficient to interpret the data — and no more. A common mistake is that figures have too many tick marks, and too many tick labels (Fig. 3). Not every tick mark needs a label — only so many as are necessary for understanding the data range. Tick marks should point outside, not into the data rectangle — at first glance we interpret everything within the data rectangle as data. Data points cannot be read with precision from figures, anyway — the function of a figure is not to show precise values, but relationships.

When designing or revising your own graphs, seek to improve the "data:ink ratio". Sometimes even *deleting ink* can increase the information content on the graph. Examples include the range-frame graphs (Fig. 4) where an axis is only drawn where there are data points along the axis, or when the value of the mean along the axis is denoted by a small gap on the axis (Fig. 4). A particularly elegant example of retaining information, but using less ink, is when a traditional box plot is replaced by the Tufte plot (Fig. 5).

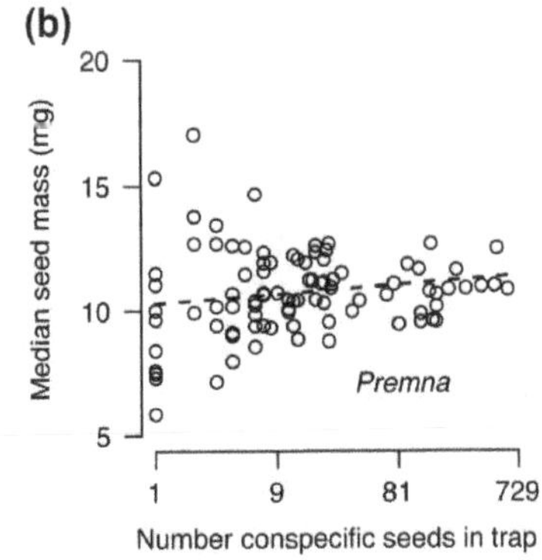

Fig. 2 Using a semi-logarithmic axis can fill the data rectangle more evenly. Here logarithmic values on base 3 was used. Published by Elsevier GmbH.

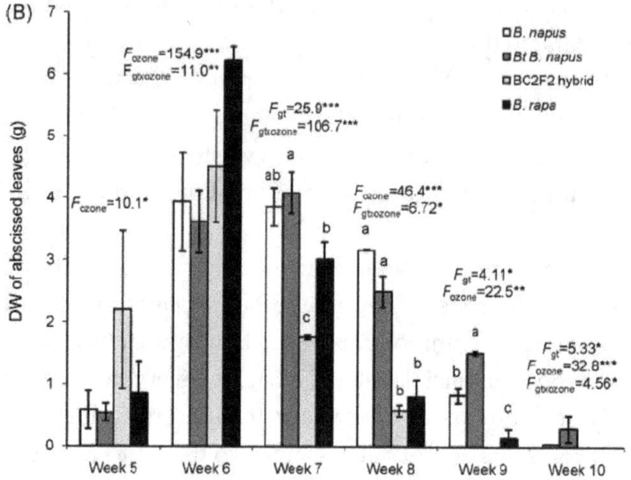

Fig. 3 Too much explanation within the data rectangle can draw the attention away from the data. Published by Elsevier GmbH.

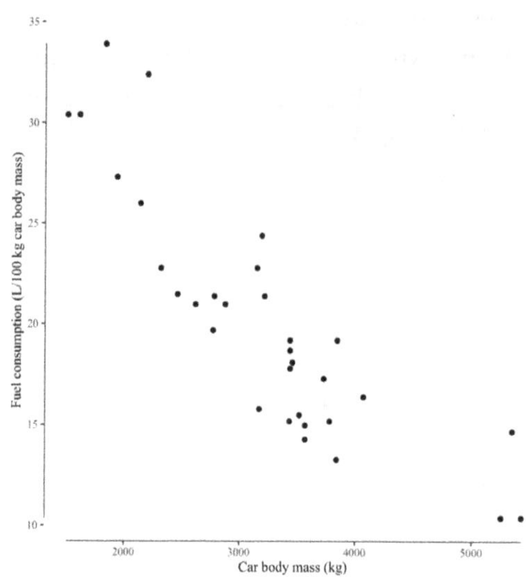

Fig. 4 A range-frame figure. Instead of the traditional axes, axis lines are only drawn within the range of data along that axis. Data modified from the mtcars R dataset, version 3.6.2. Image by author (2020).

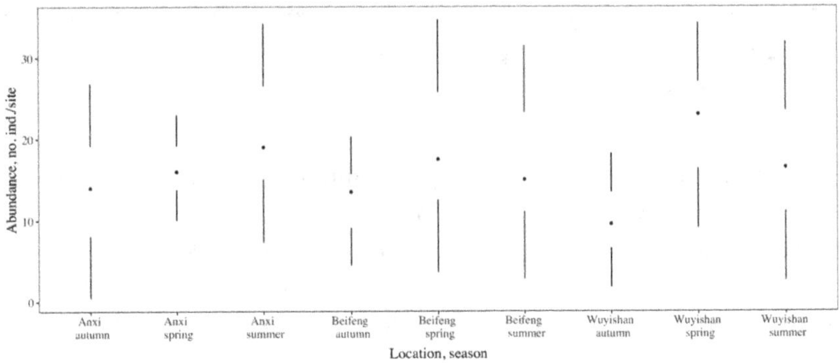

Fig. 5 A Tufte plot. An elegant way to show comparisons in descriptive statistics — in this case, the median, the central quartiles and the range.

Integrity

Integrity requires that the kind of data pictured should always be truthful and unequivocally identifiable. All axes should have an axis label, even if it seems superfluous. This should contain both what was measured (what is pictured along that axis) and the measurement unit, if there is one. Sometimes we picture dimensionless numbers or indices. These have no measurement unit.

Integrity also requires truthfulness in relation to data dimensions. Data should not be pictured with "pseudo-dimensions". A common mistake is to present two-dimensional data "in space", using a third dimension (Fig. 6). The "thickness" of the cake on Figure 6 is totally irrelevant — it carries zero information — because, by intent, the area and not the volume of the various slices are to be compared. This is called a pseudo-dimension, because the same data can be presented in one, or even two fewer dimensions without loss of information. Data dimensionality and figure dimensionality should be consistent. We struggle, in any case, to picture anything beyond three dimensions — the three that we have should be used wisely.

There is also a sensory physiological reason for not exceeding the necessary number of dimensions: the human eye is very good at perceiving even minute differences in linear dimensions. We are much poorer in distinguishing area differences, and almost hopeless when it comes to judging differences in volume. Do not use more than the necessary number of dimensions to present your data (see more about this later in this chapter).

Data differences, and their representation, should be truthful: if the difference in your data is 50%, presenting this as a 100% difference would give a false impression to the viewer. This has been called the "lie factor" (Tufte, 2006), and is defined as the ratio between the difference in representation divided by the difference in measured values.

No axis breaks are allowed. The eye instinctively connects these gaps along the minimum chord and, thus, we get a false impressions of trend lines (Fig. 7). When interpreting figures, we estimate shape, length, etc. and make comparisons. Often there are several graphs that are to be compared. The axes should be consistent among such comparable panels. We should be careful here, because primitive graphing programs present any graph panel in the same size, irrespective of the axis range.

If such consistent axes are not possible, or if they grossly violate the principle of filling the data rectangle with data, units of change should be physically identical (Fig. 8).

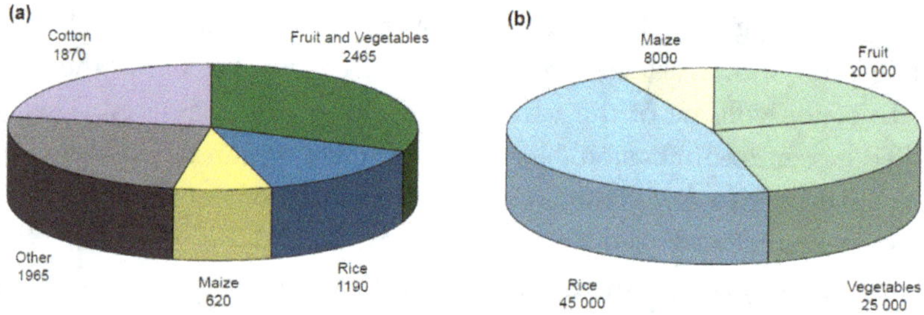

Fig. 6 An example of a superfluous third dimension, which carries no information at all — thearea of the circle diagram is related to the number represented. Notice also the double data presentation.

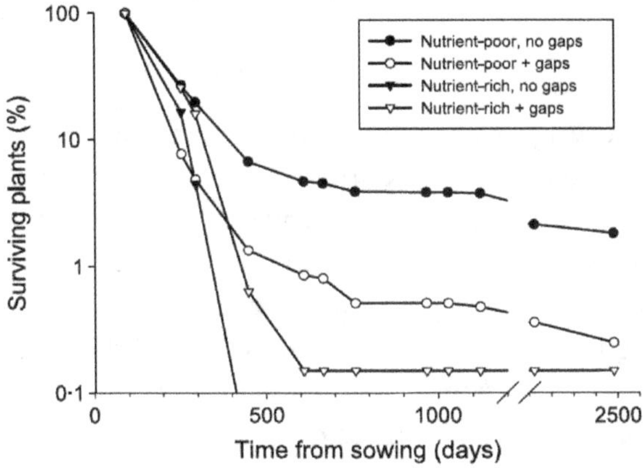

Fig. 7 When the axis is broken, we cannot correctly judge the slope of the curve.

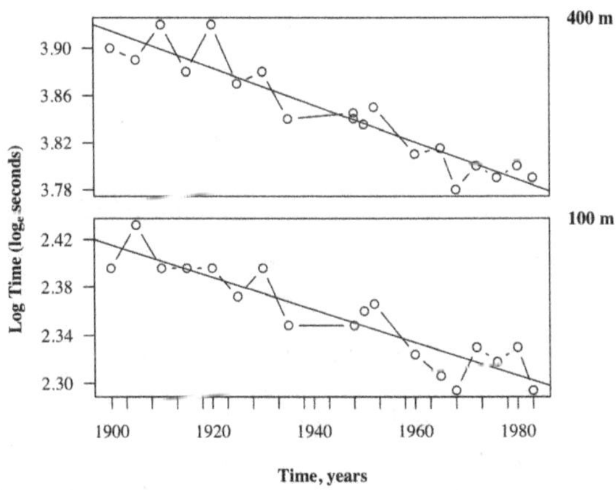

Fig. 8 When multi-panel figures are presented, their axes must be identical, or at least comparable. In this example, different sections of the same scale are used on the vertical axis. Data show the (natural logarithms of) winning times at 20th century Olympic Games at 100m and 400m sprint distances. Image by author (2020). Data modified from Cleveland (1993).

Clarity

Allow for reduction

Figures are almost never reproduced in their original size. A majority of figures is reduced when reproduced in a journal; reduction always results in loss of sharpness, definition, and detail. Consequently, figures should be designed to withstand reduction. If in doubt, use a photocopier to generate a reduced-size figure. Consider that reducing the length of axes by half would result in a figure size of 25% of the original. Also, consider the final dimensions of figures that can be reproduced in the journal. If possible, try to fit to these dimensions — but remember, the journal is the medium, and the information is the message. The medium should never take precedence over the message — if your figure must be an odd size, then so be it. You will, though, have to justify your choice of dimensions to the editor.

Data should stand out

The size of data points should always be large enough to identify where, and how many, they are. A continuous line without data points suggests an endless number of measurements, which is rarely the case.

Data visibility: symbols, overlap, axis interference

When presenting several data series on one graph, we should use symbols that easily and correctly identify data groups (see Box 10).

Box 10. Optimal symbol sets for graphs

The default symbol set should always be the empty circle. This allows the distinction between points, even if they are up to 98% overlapping. No other symbol has this advantageous feature. Be careful, because the default symbol is different in most (even scientific) graphical programs.

When different set of measurements are presented on one figure, it helps if you can connect the data points. Even in this case, however, try not to put more than 5 measurement series on one graph. Remember the 'no double coding' principle. It is usually better to use different symbols and simple connecting lines rather than the same symbol and different connecting lines.

When connecting data points does not make sense, things are becoming complicated. In such a situation, the viewer has two tasks: to be able to group data points that belong together (a unification task) but also to be able to distinguish between the different data sets (a separation task). Cleveland (1994), after thoroughly analysing such situations, suggests two symbol sets to use:

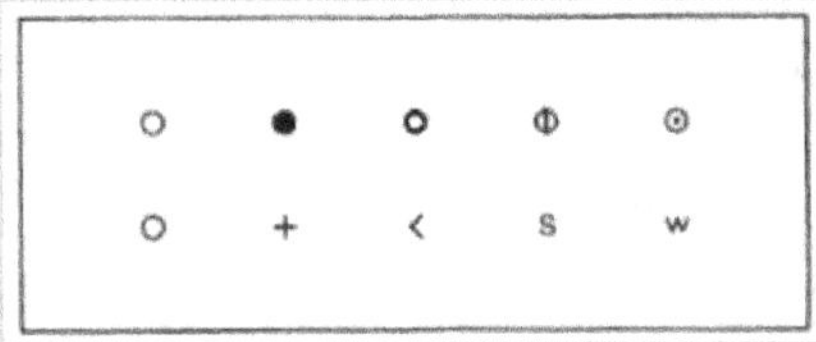

Plotting Symbols. Image from Cleveland (1994), p. 164.

When the data points do not show too much overlap, the top set of symbols gives the best possibility to recognise the data points that belong together and separate the points belonging to different data sets. When there is considerable overlap, use the bottom symbol set.

More than five sets of data can rarely be pictured within the same data rectangle. Use a multi-panel graph in such cases, possibly with a background grid to ease comparison; remember the comparable axes.

In cases where there are precisely overlapping measurements, the true impression of the distribution of the data requires that these data should be recognised as separate measurements. This requires the use of empty circles as symbols, because they can be recognised as separate even when they show >90% overlap. Writing a number by the data point is not an acceptable solution (see Fig. 8a as a bad example) as we cannot mentally transform a number into that many separate symbols. If there are only a few overlapping data points, they can be drawn close to each other, even touching (Fig. 9). The best way, however, is a process called "jittering". This process adds a small random number to the data, and the graphing program subsequently treats them as separate points. Following this method, we can obtain a reasonably correct impression of the data distribution (Fig. 10). Naturally, this method should only be used to graph your data.

Sometimes we have data sitting on the x or y axis (Fig. 11). The emphasis on the data points allows the axis to be moved away (Fig. 12). This makes the interpretation of the data easier.

The principle of clarity requires that the emphasis is on the data. The data rectangle should be filled by the data only. At first glance, we try to interpret anything in the data rectangle as data. Therefore, any additional elements, including tick marks, legend, etc. should, where possible, always be outside the data rectangle. This helps us to interpret the figure. Marks, labels, and tick marks pointing into the data rectangle may generate confusion.

Comparing different panels can be difficult, and background grid lines (Fig. 13) and complementary marks can be useful. Such marks should be clearly different (lighter) from data and lines so as not to confuse the reader.

Data are interesting. People designing "interactive", "interesting" data presentation methods try to move the emphasis from content to form — a bad design principle. Such practice often tries to disguise shallow thinking and/or an underestimation of the viewer. If you believe do not have interesting data, the wrong type of data was measured, and no amount of decoration will make them interesting.

Graphs with two dimensions can be wider than tall, as well as taller than wide. Which one is better? Our eyes are instinctively used to scanning horizontally, so we find the wider-than-tall shape easier to interpret. This is especially so if we have a data with a lot of variability, creating a curve that is "wiggly". We can make a curve easier to interpret if the diagonal sections are as close to 45 degrees as possible. This is useful because our eyes can also detect small deviations from the diagonal. This technique is called "banking to 45 degrees", and is automatically generated by more sophisticated graphing software. This is done iteratively by changing the height and width of a graph until most of the diagonal lines are as close to 45 forty-five degrees as possible.

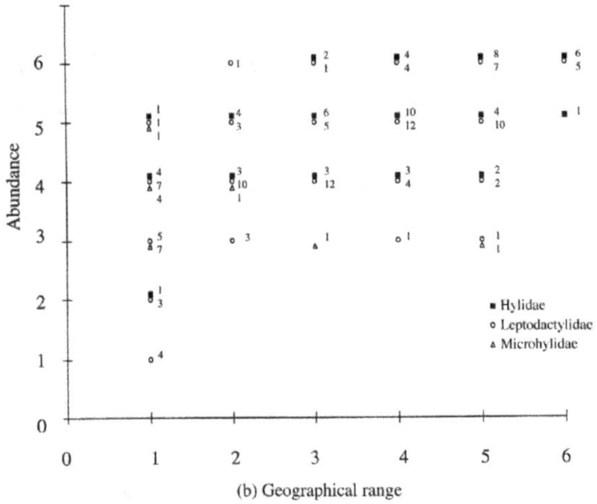

Fig. 8a When identical data points are to be represented, do not combine two forms of data presentation as here (symbols accompanied by numbers).

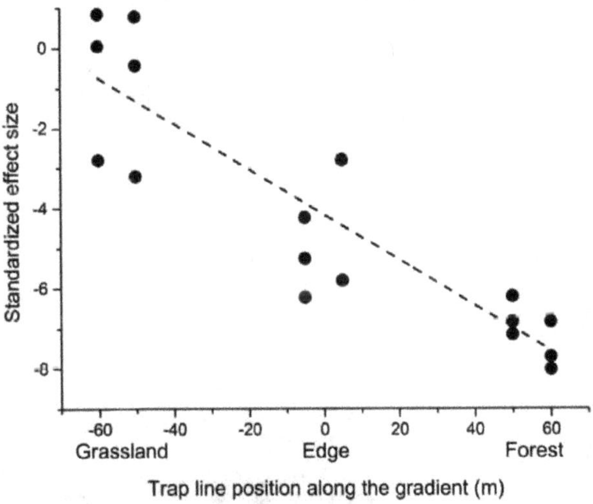

Fig. 9 When there are few identical data points, you can draw them touching each other, as the two data points here at x=50 and 60. The resulting distortion is not large.

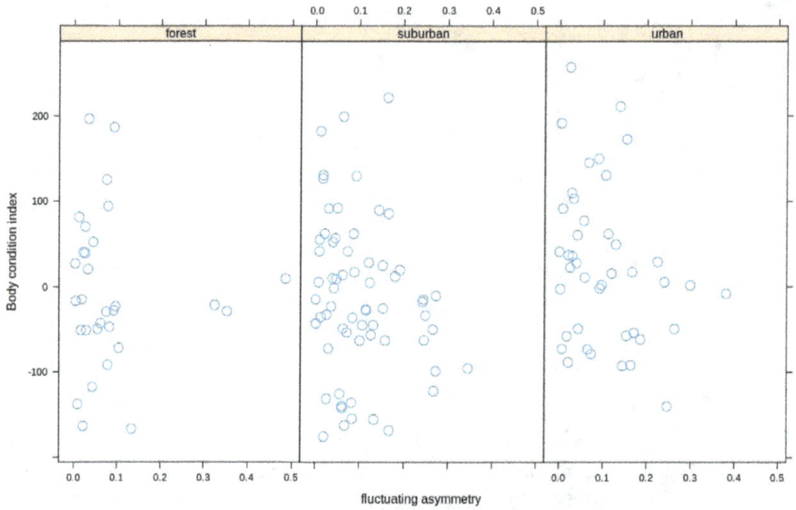

Fig. 10 When numerous data points overlap, drawing them as touching points would grossly distort graphical perception. In such cases, a judicious degree of jittering allows the viewer to distinguish between identical points with minimum distortion in trend perception.

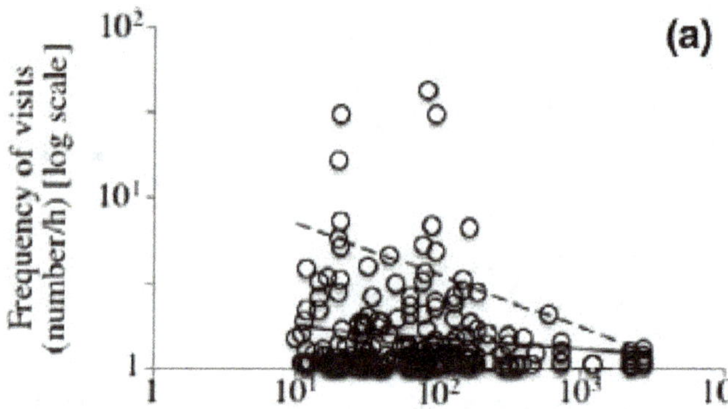

Fig. 11 Too many data points sitting on the horizontal axis make this graph cluttered.

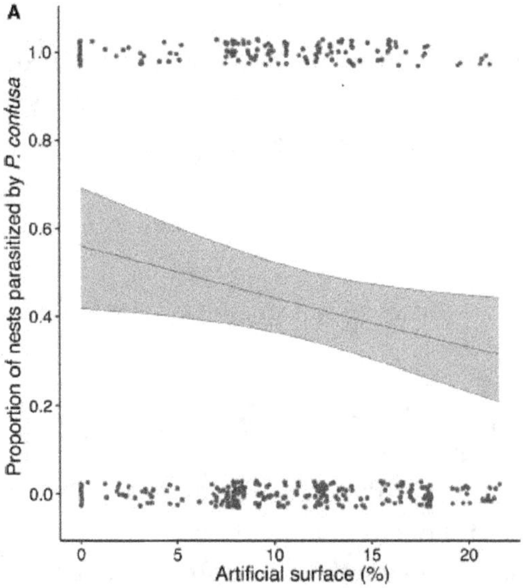

Fig. 12 Moving away the horizontal axis increases clarity. Note, however, that full dots are suboptimal symbols — close-lying data points are not easy to distinguish.

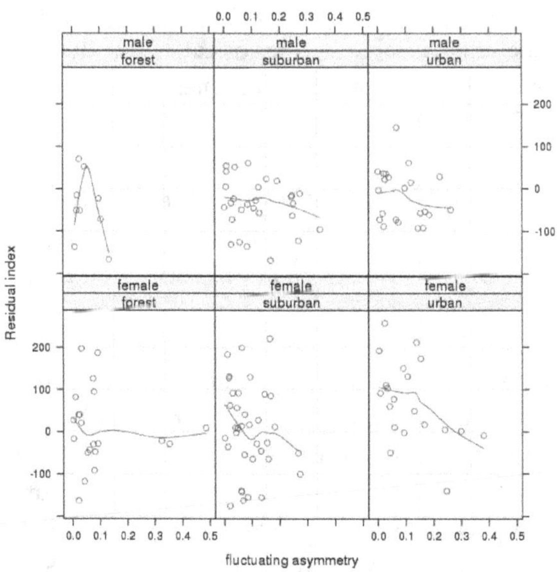

Fig. 13 An identical, visually gentle background grid helps to compare the position of the data points on different panels.

Clutter

Clutter is generated when different elements of text, of various size and font type, are used in the graph. Often, there is simply too much explanation (Fig. 3), drawing attention away from the data points themselves.

Graph Types to Avoid

Bar chart

The bar chart is one of the oldest types of scientific figures (Playfair, 1622), and it is still one of the main types of figure used in published articles. The data value is represented by the height of a column — the width of the column is irrelevant. Note that there is a difference between a histogram and a bar chart, even if they seem superficially similar — in the former the width has a meaning. Bar charts can often be replaced by a simple dot, appropriately placed. An additional problem that many programs create is that, on the horizontal axis of a bar chart, the axis variable is categorical — but the program might arrange them at even distances, sometimes falsifying the time trend (Fig. 14).

Bar charts should be avoided because they use two dimensions to present one-dimensional data (the bar has a height and a width — but only the former carries any information), thus violating the principle of economy.

When bars are grouped into a composite bar chart, comparison between them becomes complicated and it is not possible to clearly perceive what they are telling us (Fig. 15). In the case of stacked bar charts (Fig. 16), comparison becomes even more is difficult because the different segments start, as well as end, at various heights, and we cannot judge the lengths correctly. Both should be avoided.

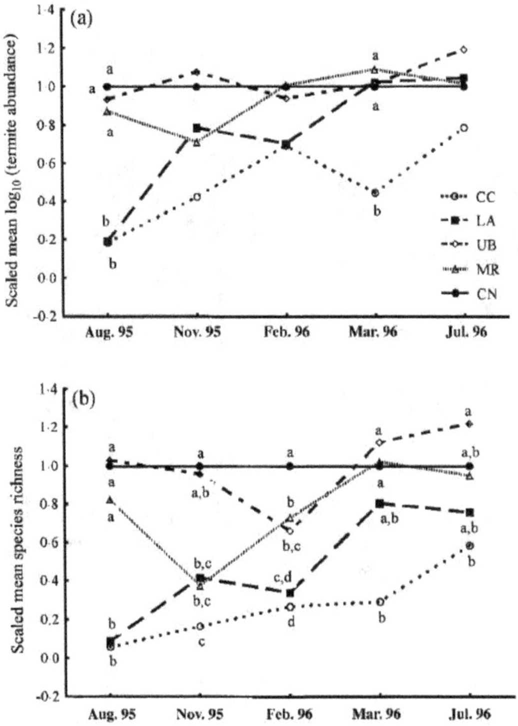

Fig. 14 Due to the incorrect scale on the horizontal axis, the rate of change cannot be correctly perceived.

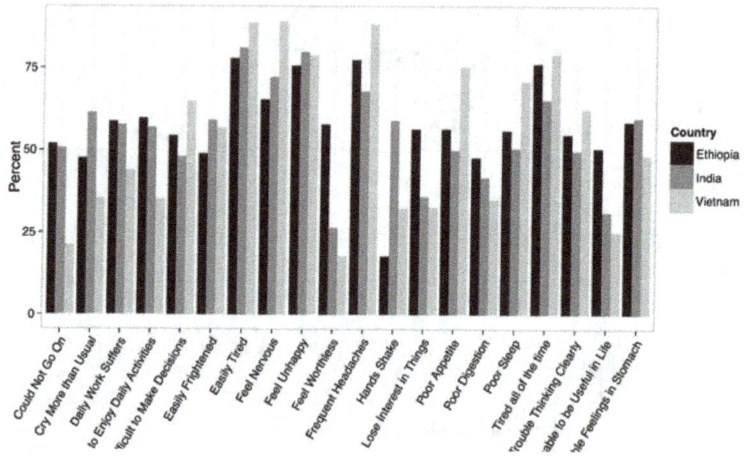

Fig. 15 A grouped bar chart. The same type of data from the three countries can be easily compared but a "country profile" is nearly impossible to perceive. Note also that the horizontal organisation, alphabetical by labels, is suboptimal.

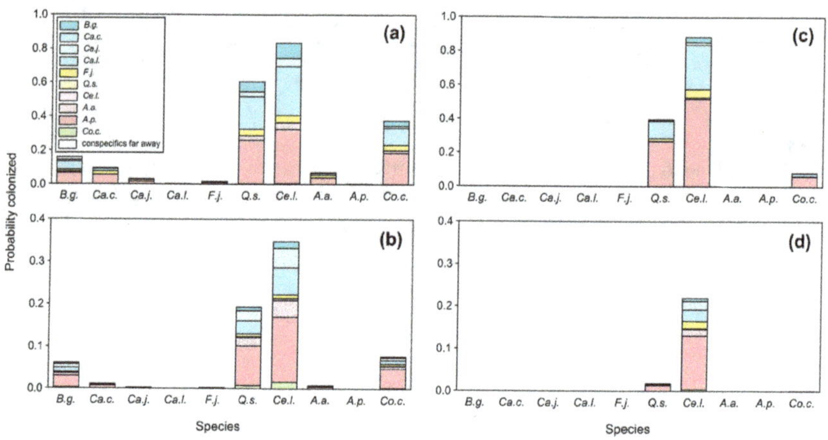

Fig. 16 A stacked bar chart. Due to the constantly shifting baselines, the smaller differences are difficult to interpret.

Pie chart

Pie charts are frequently used in business and on "pseudo-scientific" graphs. This is a bad design overall, because we are asked to make judgements based on the area occupied by the segments. Our eye cannot judge differences in area well, and only the biggest differences can be identified correctly. That renders this figure type virtually useless. This fact is inadvertently admitted by the common practice of dividing the pie into a few sections only, as well as by writing the actual values of the pie chart segments near the segments themselves (and the common program option that allows one to generate such figures) — thus also breaking the principle of "no double data presentation" (Fig. 17).

There are now several new methods of data presentation available, developed during the last 30 years. Many of these were pioneered by the team at Bell Laboratories, with the leadership of William Cleveland.

The preferred graph to present one-dimensional data should be the dot plot. The dot plot is a relatively new graphical method, even if it is beguilingly simple — the first dot plot was apparently published in 1984 (Cleveland, 1993). This is eminently suitable to present one-dimensional data, where other variables are categorical ones. Traditionally, the measured variable is pictured along the horizontal axis, and the labels are outside the panel, along the vertical axis (Fig. 18). If there is more than one categorical variable, the multi-way dot plot can be used (Fig. 19). In such cases, carefully consider the primary vs. secondary comparisons (Fig. 20).

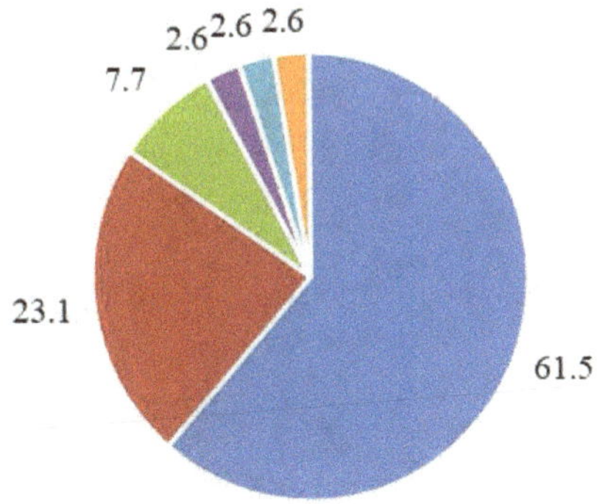

Fig. 17 A circle diagram with numbers: a case of double data presentation.

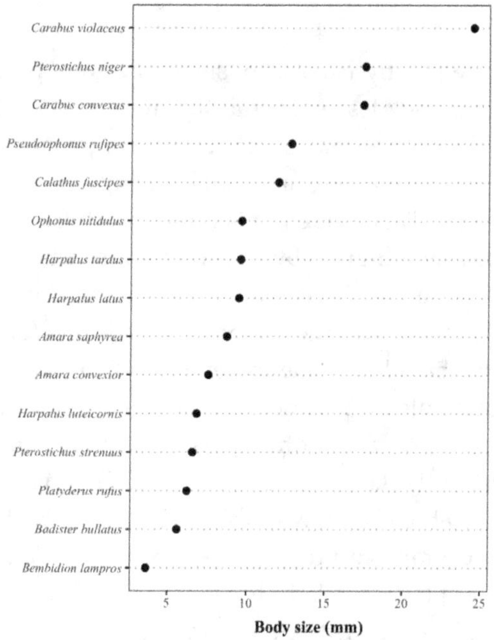

Fig. 18 A sample dot plot. Differences are presented along one dimension, e.g. length. This allows a precise perception of pattern as well as individal differences, even small ones. M. Ferrante, unpublished data. Image by author (2020).

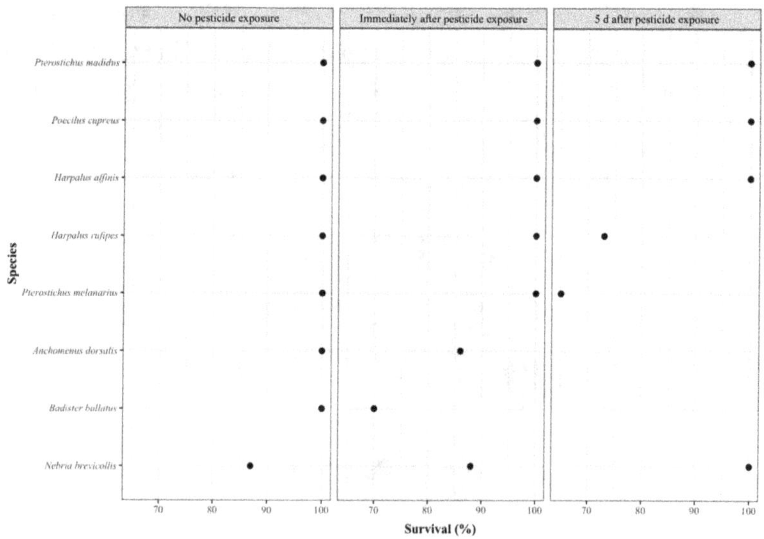

Fig. 19 A multi-way dot plot, presenting survival data by various ground beetle species exposed to pesticides. The emphasis here is on the effect of treatments on the different species. Data from Greenop et al. (2020). Image by author (2020).

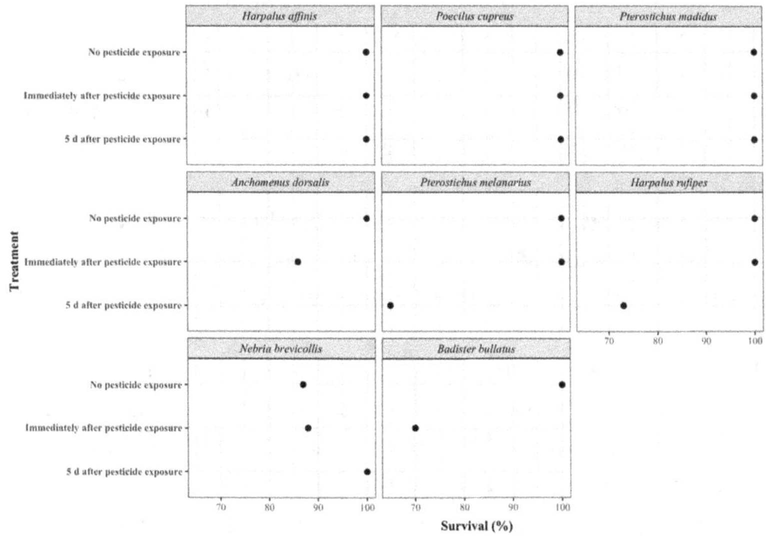

Fig. 20 An alternative multi-way dot plot of the same data as on Figure 19. Now the reaction by the different ground beetle species to the treatments is the focal comparison. Data as for Figure 19. Image by author (2020).

Within the panel, the largest values should be near the top of the panel, sequentially followed by smaller ones. This is useful for Gestalt (shape) perception. In multi-way dot plots, an optimisation algorithm should be employed, so that the larger values appear most frequently near the top (Cleveland, 1994).

Two-dimensional data are the most common type, and a line-and-symbol plot is suitable for presenting such data. There are several additional methods, especially useful for data exploration, including the locally weighted regression (loess or lowess), the conditional plot, or co-plot, and the scatterplot matrix. These and other useful graphical methods can be found in Cleveland (1993, 1994).

Figures in papers vs. figures used in talks

Never use a figure published in a paper as illustration in an oral presentation. The reason is that the first one will be printed, and readers can absorb and evaluate it at their own pace. It will not vanish. When a figure is used in a talk, it must be understood over a shorter period of time, set by the speaker — the figure is only visible while it is projected during the talk. Consequently, these two types of figures are not interchangeable: a good figure for a paper usually contains too much information to be useful in a talk. In the same way, a figure that is useful in a talk rarely has sufficient information to be worthwhile as a figure in a paper. Complex information can be conveyed by a single, carefully designed figure. The same information needs several figures in a talk.

Furthermore it is possible, and advisable, to use colour in a figure used in a talk, while the use of colour is usually prohibitively expensive to publish in a paper. Figures in a paper should usually be black-and-white.

Therefore, prepare different, specially tailored figures for a paper and a talk, even if the same data are to be presented. When thinking of figures that will be published in a paper, think black and white. Consider the careful use of colour for figures to be projected during a talk. Internet publishing often allows authors to publish colour figures at no extra cost. However, the above constraints on complexity vs. time allocated to interpret figures mean we should still not use a published colour figure in an oral presentation.

The same goes for a poster, mainly due to the type of setting in which a poster is typically presented. At a conference, too many posters usually compete for the participants' attention, and they cover too many different things. In theory, viewers can spend unlimited time in front of a poster but, in reality, they rarely do so. Therefore, a poster is closer to a talk than to a paper.

Colour

The use of colour used to be rare in scientific journals. The reason was cost: printing black-and-white text, figures and photos is more expensive than printing text, but much cheaper than printing the same in colour, mostly because better quality paper and more sophisticated printing machines had to be used. Moreover, printing is done on sheets of paper, usually equivalent to 16 pages (these are cut up after printing). Consequently, those 16 pages all had to be printed using the same machine, even for only a single colour figure. The extra cost, which was nearly always passed on to the author, was usually over USD 1,000. This made colour illustrations very rare in the scientific journals. This is changing: many journals publish colour in their digital issues for free, while they still print the same figures in black-and-white in the printed edition. However, this is becoming less and less justifiable, as the cost of printing in colour vs. black-and-white is no longer so different.

Use colour to help understanding, not for decoration. Modest use of colour is very helpful, but do not abuse the technical possibilities to generate many different colours. Try to use harmonious combinations for maximum contrast. These are formed by colours at opposite positions on the colour wheel. Differences in shade can work, but they are very much equipment-dependent. What appears a nice shade on your own screen, could turn to a garish colour on a projector, if driven by a different computer.

There are several helpful programs on the Internet to help choose appropriate colour combinations. The Colour Scheme Designer (www.colorschemedesigner.com, now migrated to www.paletton.com) also allows you to check how people with various colour-related vision impairments will see your chosen colour scheme.

Proportion, Scale and Appearance

Graphs should tend towards horizontal, being wider than tall. Our eye is naturally practised in detecting deviations from the horizon. This also helps with ease of labelling. Avoid labels aligned in various directions. In keeping with reading direction (left to right), we can look up the impact of the cause (independent variable) on the effect (dependent variable) along the horizontal axis. When the cause is presented on the horizontal axis, you have horizontal depth, i.e. space to elaborate.

Box 11. Reviewing/evaluating figures

1. The first question must be: Do the data justify a figure? Should these data be put into a table, or can they be written in the text?

2. Is the type of figure acceptable? Is a better type of figure necessary? (dot plot, multiple dot plots, co-plot, scatterplot vs. histogram or pie chart)

Economy:

How is the data/ink ratio? Can this be improved?

Is the data rectangle well filled with data? Are any elements that do not belong there?

Integrity:

Is data dimensionality observed? Are the axes appropriate? Is context provided? If multi-panel figure, is the comparability of panels observed?

Clarity:

Are the axis scales appropriate? Are the ticks, tick labels, and symbols OK? Are both axes clearly labelled, with units of measurements? Can ink be eliminated and information retained?

Format: is the size of the figure, the font size and type, the width: height ratio of the figure acceptable?

Is banking to 45 ° to be employed? Is there no superficial decoration, unnecessary visual novelty? Are the symbols well chosen and visible? Are there overlapping symbols, or symbols sitting on axes — if so, are the axes moved to make these symbols visible? Does the figure withstand reduction when printed?

Photographs

Photographs can be a useful addition to your paper. However, they are not decoration — they should be informative. Photos are more expensive than text to produce, so editors are vigilant when it comes to including photographs. The editor will almost always ask if you really do need a photo — so you had better have a reason to justify your request. Photos of habitats, organisms, occasionally of equipment can be useful. In some cases, a microphotograph, a gel photo, or a crystallogram is the vital piece of evidence, and it should be included.

If you decide your paper needs a photo, carefully check the reproduction standard of the chosen journal. Only high-quality photos are worth including — but the printing quality will also have to be high. Colour photos are nearly always at your expense, and they can be very costly.

Black-and-white photos usually do not cost you extra. Many journals have both electronic and printed versions, and such journals often allow you to include a colour photo, at no extra expense, in the electronic version of your paper, while the printed version will carry a black-and-white photo. In that case, it is best to submit two versions of the photo; one in colour, and the other one in black and white.

As mentioned earlier, figures are almost inevitably reduced in size when printed. This always causes a loss of detail and definition. You can plan for this in the relatively simple line graphs (although it does need attention), but such diminishment may be fatal to photos: so how can you control photo quality? The best result is achieved if there is no reduction or enlargement. It is very rare that every square centimetre of your photo is vital, so experiment with cropping, i.e. selecting the part of the photo that is important (this is often only a small part of the photo). Many photographic programs enable this. Suitably cropped photos can be reproduced without reduction, thus retaining the original level of detail, but it is useful to consider the dimensions of the journal. If you manipulate your photo electronically, this *must* be declared on the caption.

When multiple photos are presented, they can be compiled into a set of images (a plate). If appropriate, you can place a scale on the photographs. If you choose to label your photos with letters, make sure these are visible. Indicate the magnification, too, in the caption.

Today, most journals work with digital photos. This part of the process is, as yet, rather shaky due to the occasional transferability problems between programs and figure formats and, thus, there are very detailed instructions on how to prepare and submit a digital photograph. Read these instructions very carefully. Check acceptable or preferred file formats; you can contact the technical editor for clarification. This will be seen as co-operation, not hindrance. This, however, is something to consider at the stage when your manuscript has been accepted for publication.

If you are asked to provide hard copies of your photos, write appropriate information on the back of the photo, in soft pencil. This in-photo information should indicate the manuscript number, the photo number, and the orientation. Do not think it is obvious — the printer is not a scientist, so do not expect her to be able to interpret the photo. Indicate the desired position of the photo in the text.

Do not forget about pen and ink illustrations. They can be very useful, but must be drawn by a professional illustrator. Use them even if you have to pay for the drawing out of your own pocket. It does not cost the earth, and you will not regret the expense.

16

Studying the Sample Graphs

It may sound surprising but, once you develop an eye for good graphs, you will notice the occasional mistake in graphing practice. Several scientific editors agree. Per Enckell, the then-editor of one of the prime journals in the field of ecology, *Oikos*, re-published in the journal a chapter from the book by Edward Tufte, the eminent practitioner of presenting visual information (Tufte, 1990). Alas, this gesture did not revolutionise graphing practice. Valiela (2001) has also devoted a chapter to provide suggestions for designing scientific graphs. He selected published graphs, analysed their imperfections, and suggested improvements. The same approach is followed in this chapter. Below, you will find some of the most frequent mistakes: too few data to merit a figure (example 1), problems with the integrity of the figure, making appropriate comparisons difficult (example 2), inappropriate coding that impedes understanding (examples 3 and 5), and clutter (example 4). In each case, I present a more acceptable version of the same figure. Box 11 also lists some of the criteria to consider when designing your graph.

Example 1. The Graph that Need Not Exist

This graph presents the survival probabilities of elephant seal juveniles (1-3 years old) and adults (4 years or older) on two sub-Antarctic islands, Macquarie and Marion (Fig. 21; Fig. 7 in McMahon et al., 2003).

There are several deficiencies in the figure concerning economy, integrity and clarity:

- the figure is too wide. It can easily be made narrower to fit into one column, saving ca. 50% of space.
- the data rectangle is not filled by data — a large part on the left is only there to accommodate the legend. This is needless — legends should be placed above the figure, so that the figure is not wider than necessary. There is no need to abbreviate "Macquarie Island". The frame around the legend is superfluous. If the legend were above the figure, the vertical axis could start at p = 0.72, saving more space.
- there is double data presentation: mortality values are represented by symbols, but precise values are also written on the figure. This is wrong, as the same data cannot be presented twice.
- there is also double coding: not only are the symbols different but the vertical lines marking the confidence intervals are all different
- the axis labels are not appropriate or are completely absent. The vertical axis should probably be "Estimated survival, %" or "Estimated probability of survival"
- the labels of the two groups, juveniles and adults, do not line up with their respective data points
- the symbols are too small — and hardly visible. The tick points are inside, whilst secondary ticks would be helpful to allow the readers to make a better estimate of the values. The lettering on the figure is a little too small, which decreases readability.

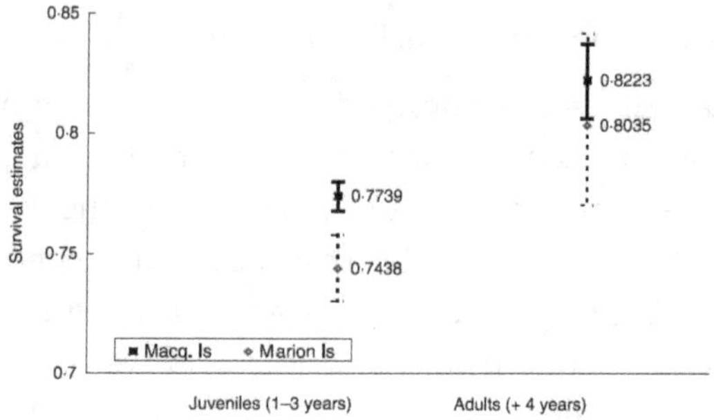

Fig. 21 A figure with too few data points. From McMahon et al. (2003).

Overall, however, the main problem is that this figure contains too little information: four data points and their relative 95% confidence intervals. The same information can be presented more economically in the text, therefore, there is no reason to construct a figure here.

Example 2. Small Effect, Big Effect: Misleading the Reader

This example is from a study on the winter mortality of Redshanks (*Tringa totanus*), in an area of Great Britain where there is a large population of wintering birds that are harassed by European Sparrowhawks (*Accipiter nisus*) preying on them (Fig. 22, Fig. 3 in Whitfield, 2003).

The first impression from the figure is that bigger flocks suffer higher winter mortality, and this relationship is steeper in the case of adults than juveniles. A close analysis of this figure shows, however, that this conclusion is not necessarily correct. The distortion arises because the physical size of the two panels are identical, yet their vertical axis scale is drastically different: on the top panel, presenting data on juvenile mortality, the range is from about 19% to 58%, while the lower panel, with the adult mortality data, ranges from about 3% to 18%. Considering the three graphing principles, other imperfections arise:

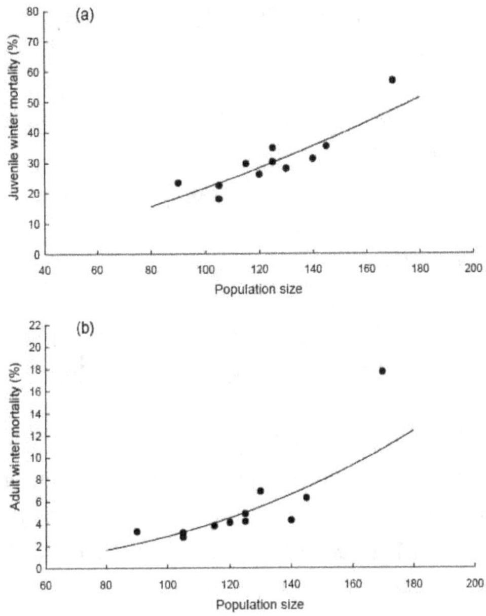

Fig. 22 A figure with two incomparable panels.

From the point of view of *economy*:

- uneconomical use of space, due to bad choice of axis intervals. There are no measurements below 90 on the horizontal axis, and nothing above 170. On the vertical axis of the upper panel, no values appear below 15 and above 60, while the range of the y values on the lower panel are from 2.5 to 18. This also forced the typesetter to place the figure between two columns, thus wasting even more space. Modifying the axes would save a lot of space;
- the full circles are not the best symbols, due to overlap on the lower panel;
- using range-frames can save some ink;
- there are more than the necessary number of tick labels on the vertical axis, especially on the lower panel;

- the vertical axis labels allow the precise identification of the data on the panel, thus the panel labels (the letters a and b) are not needed.

Concerning *integrity*, the two panels are not comparable: they present very different ranges, yet their physical size is identical. This misleads the reader, who is not able to correctly interpret the relationship between flock size and mortality in adults vs. juveniles.

From the point of the third principle, *clarity*:

- the ticks point inside, into the data rectangle. The data rectangle should be reserved for data only;
- the symbol sizes, axis labels, and tick labels are all a little too small, just about readable in the original paper. Reducing the number of tick labels not only improves the economy of the figure, but would also allow an increase in the size of the tick labels, making them more readable;
- there are no measurement units on the horizontal axis label. It probably should read "Population size, no. of individuals".

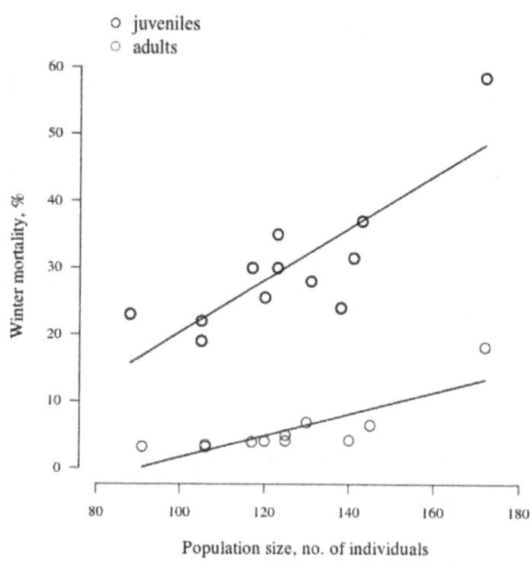

Fig. 23 Data from Figure 22, re-drawn. The two data series are now straightforwardly comparable, and the trends can be truthfully interpreted. Image by author (2020).

On the redesigned figure (Fig. 23), both the measurement series appear on the same graph, because they do not overlap. Now it is obvious that both fit a linear regression, and the relationship between mortality and population size is steeper in the case of juveniles compared to adults. The axis intervals are reduced, and the two axes do not touch. Empty circles are used as symbols, with increased sizes; letter size is increased, and a serif font is used, which has better readability than the original sans serif. A small simplification is that the axis label units are not in parentheses, but are separated from the measured parameter by a simple comma.

The highest mortality values in both age classes appear in the largest population. The difference is so great that these almost seem like outliers. Due to the lack of more data points from larger populations, however, the suspicion must remain unconfirmed that there may be a threshold size over which predation pressure radically increases.

Example 3. The Chaotic Figure — Coding Can Stand in the Way of Understanding

This paper reports on the effect on termites of various understory treatments in a tropical forest (Davies et al., 1999), from complete clearance to selective clearing up of termite mounds and undergrowth. Figure 24 (Fig. 2 of the original paper) presents the changes in density and species richness of termites over a one year period, considering the changes in the untreated control area as baseline. The trends are not simple, the lines criss-cross each other, but the figure design does not allow fast and effective decoding.

The biggest mistake concerning the *economy* of this figure is the double coding: the treatments are marked by different symbols; in addition, they are connected by different lines. This is needless and breaks the "no double coding" principle. The figure has more visual novelty than necessary for swift decoding.

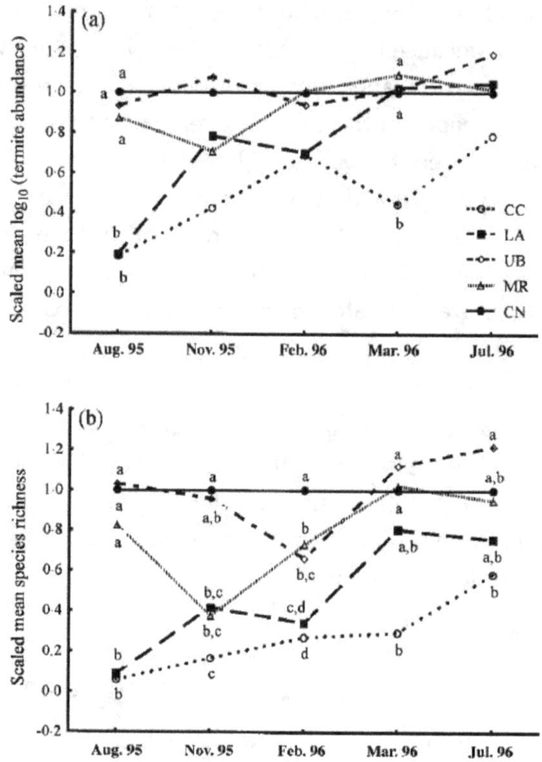

Fig. 24 A cluttered figure with faulty horizontal axis.

The axis lengths are also longer than necessary, probably to accommodate the legend which should be outside the data rectangle, anyway; the vertical axis starts at -0.2, (which is an impossible value in this context). Thus, the figure uses more space than justified. The horizontal axis is also longer than optimal: it starts before the first, and runs after the last sampling occasion, making the figure wider than needed. Consequently, the printer was forced to use the full width of the page to place the figure, creating large empty spaces on the page on both sides.

From the points of *integrity*, there are two points to mention:

- the horizontal scale is wrong: the equidistantly marked sampling occasions suggest that they were taken at equal time intervals. The period between two sampling occasions varies from one month (February-March, 1996) to four months (March — July, 1996). Consequently, the figure distorts the time trend, misleading the reader;

- measures of variance are missing — as the values are means, according to the axis labels, some measure of variance is necessary. There are labels on the graph, indicating the significance of differences between data points, but this is neither complete, nor easy to interpret.

The *clarity* of the figure is also suboptimal:

- ticks point inside the data rectangle. Due to the larger-than-necessary area occupied by the figure, these do not obscure data, but this may change when the data rectangle is reduced to its necessary minimum size. There are too many tick labels on the vertical axis, while the horizontal axis label is missing. The symbol sizes are too small, and not easily distinguishable. This is partly due to symbol choice — several of them are too similar to allow fast and precise identification of the individual treatments. The legend is placed opportunistically inside the data rectangle, but the data rectangle had to be increased, otherwise the legend would not fit. The codes are cryptic, and their meaning is not given in the caption, either. This breaks the principle that a figure, in combination with its caption, must be understandable without reference to other parts of the article;

- the reference line appears with codes that serve no clear purpose. The codes for statistical comparisons are also complicated.

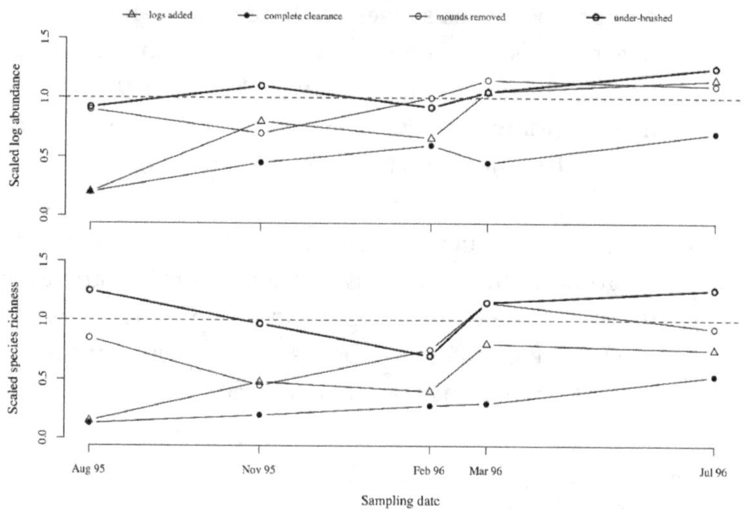

Fig. 25 The horizontal axis with true-to-time indicates a sudden change in some sites between February and March. Image by author (2020).

This figure can benefit from a range of improvements. The treatment coding can be written out, instead of using cryptic abbreviations. The legend can be placed above the data rectangle, reducing clutter. The control is coded simply as a horizontal line. The horizontal axis is modified to indicate correctly the dates when samples were taken, thus depicting the proper time trend. Now, the sudden change between February and March is very clear, hinting at an important seasonal factor change. The double coding is eliminated. The curves and lines are simplified. The two panels have the same vertical scale. The size of the symbols is increased; the font is larger and is changed to a serif font for easier readability. The axis labels are modified to indicate the parameter as well as the measurement unit. The tick labels are fewer and point outside. The coding of the differences between treatments is also simplified: letter codes appear only by those data points that are significantly different from the reference (control) value.

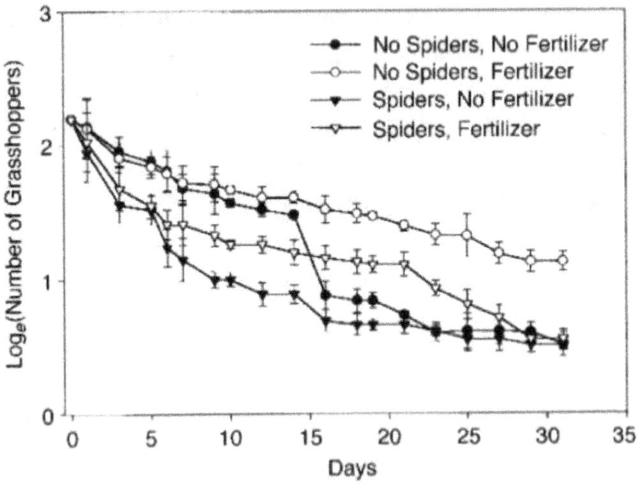

Fig. 26 A figure with a seemingly unavoidable clutter.

Example 4. Reducing Clutter

We often study an effect that unfolds over time. In such cases the starting conditions should be identical, so that the effects of the different treatments are comparable. Once the treatment starts to affect the response variable, differences appear and gradually become larger — but in the early phases of the experiment, the differences are small. Presenting such data using a figure poses a special problem, because overlap between the different values/curves is almost inevitable. Such is the case in the experiment reported by Oedekoven and Joern (2000), who examined the effect of spiders and use of fertiliser on grasshopper density on host plants. However, the resulting figure — Figure 3 in the original, here Figure 26 — contains several problems.

From the point of *economy*, the figure occupies more space than is justified. To reduce overlap among the response curves, the figure is wider than it is tall. Unfortunately, this graph was designed without considering the page size of the publishing journal: the figure is wider than one column, but not quite wide enough to span the whole width of the page. A large, ugly, empty space remains, and the typographer selected an unfortunate solution: placing the caption in the middle of the space to the left of the data rectangle. The data rectangle is also larger than necessary, to accommodate the legend and due to the excessive vertical scale. The vertical scale could start at 0.5 and end at 2.5; in which case, the data rectangle would be smaller and better filled, but, apparently, the graphing program did not allow such scaling. Evidently, this deficiency was not judged an important enough issue, by either the authors, or the journal editors, to be corrected.

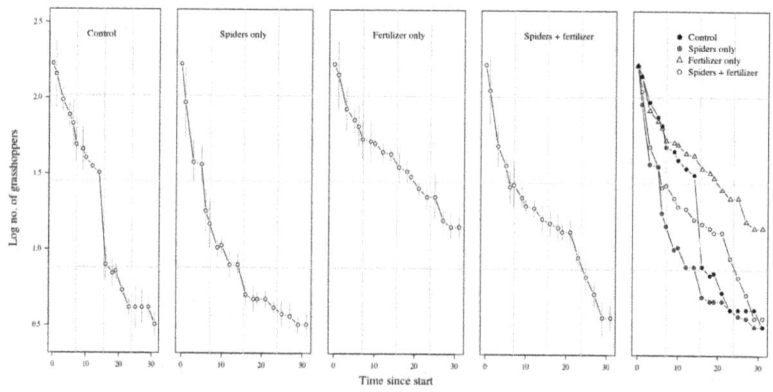

Fig. 27 The clutter on Figure 26 can be removed by plotting individual treatments separately. The background grid makes the different panels easily comparable. Image by author (2020).

The *clarity* of this figure is hampered mainly by the large overlap of the different curves, especially in the early days of the experiment but, also, near the end, where the curves again start to converge. Cleveland (1994) suggested that although, in general, figures should be wider than tall, when the curve is not very "wiggly", the figure can be made taller than wide.

This figure can be improved by presenting the responses of grasshoppers to the individual treatments on separate panels, where the variability data are clearly visible (Fig. 27). Panel sizes have been reduced, achieving better economy. The same symbol type can now be used on the various panels, and the common background grid helps the between-panel comparisons. Comparison is further eased by the fifth panel, where only the means are presented, without the variability data and, thus, without clutter.

Example 5. Complicated Coding Hinders Interpretation

Cho and Lee (2019) analysed the microbiota in three species of Arctic birds: the Pink-footed goose (*Anser brachyrhynchus*), the Sanderling (*Calidris alba*) and the Snow bunting (*Plectrophenax nivalis*). Their figure 3 presents information on the relative abundances of dominant bacterial phyla identified in the digestive tract of a few individuals belonging to the three species (Fig. 28). This is on a divided bar chart, expressed as a percentage of relative abundance per phylum.

The bar chart is the oldest documented type of scientific graph, first used by Scottish economist William Playright (Tufte, 2001). Since then, it has remained a frequently used type of scientific graph, yet it is not always the best one to present complex data; this example displays several of the disadvantages.

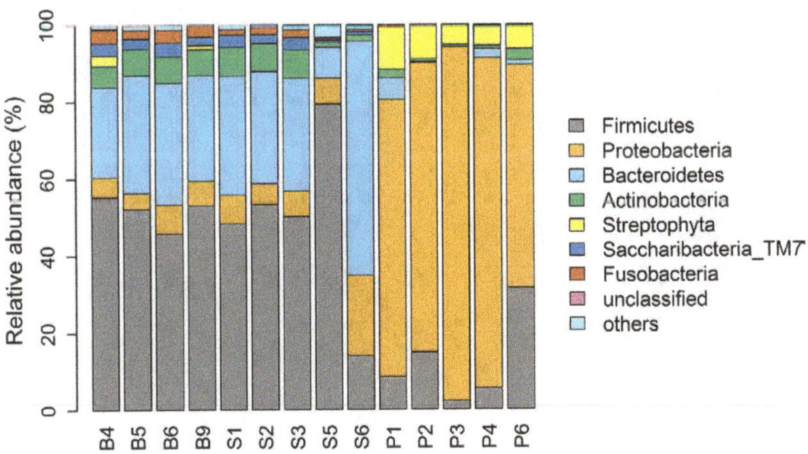

Fig. 28 Relative abundances of dominant bacterial phyla in the faeces of three arctic birds (B4, B5, B6, B9-Snow bunting; S1, S2, S3, S5, S6-Sanderling; P1, P2, P3, P4, P6-Pink-footed goose). The divided bar chart does not allow the reader to interpret small differences; only the big trends are decipherable.

This dataset is one-dimensional: there are three variables, two of which are nominal (taxonomic names or bird identities). Only the percentage of relative abundance is a measured variable (to which a measurement unit can be attached, which is %). To use two dimensions to present the data, as on the original figure, is an abuse of dimensions — and, here, as in most cases, the width of the individual columns is indifferent — the second dimension, the column width, carries zero information. Only the length/height of different segments is important. Interpretation is made near-impossible by the continuously shifting baselines: many segments start at various positions within the columns, and this lack of a common baseline allows only crude comparisons. The colour coding does not help — several of the shades used are not easy to separate. In short, we are unable to perceive any pattern except the crudest differences — for which one does not need a figure. A different type of figure is called for.

A better method to present such data is the multi-way dot plot (Fig. 29). This allows the clear coding and separation of the various categories of endangerment of the groups of shrimps, and the measured variable is presented by a linear length with a common (vertical) baseline. In such cases, Gestalt perception is also easy. There is no need to use colour, and within-panel comparisons are swift and precise.

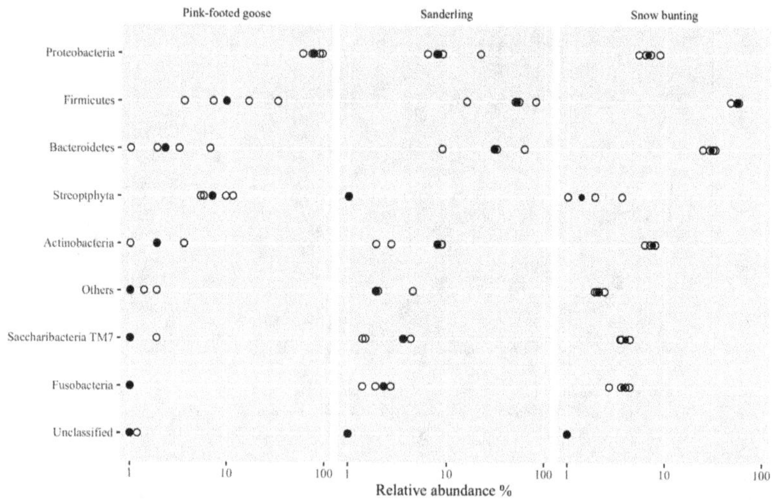

Fig. 29 A multi-way dot plot makes Gestalt recognition possible, as well as comparing microbiome profiles within and between the studied species. Data from Cho & Lee (2020) redrawn. Note the logarithmic scale on the horizontal axis. The empty circles indicate samples from individual birds; filled dots indicate the means. Image by author (2020).

17

Tables: Design and Structure

The first question is: Do you need a table? Tables are useful if repetitive data must be presented, and the precise values have importance. However, it is not good science to publish data just because you measured them. Printing a table is costly, and editors and reviewers scrutinise tables closely. Tables that have lots of standard conditions, lots of 0s, 100% or +/- s, or word lists, are usually not necessary.

Tables, just as figures, must also be self-explanatory: collectively, the title, table headings, and footnotes must allow the reader to understand the content of the table, without reference to the text.

Tables have a special format. The body of a table is organised into columns. There are no vertical lines, and only a few of the horizontal ones (see Box 12). The title of the table is always above the table itself. A horizontal line under this title separates it from the next element, the column heading. Below this, the body of the table is also separated from the parts above by a horizontal line. The end of the table is also marked by a horizontal line, separating the footnotes (if present) from the body of the table.

In the table heading, there can be partial horizontal lines, indicating sub-grouping. The extension of the horizontal line over certain columns indicates where the information above the line extends.

The organisation of the table is different from that of text. Table "lookup" is vertical, not horizontal: when we try to interpret a table, we look at columns first, so the numbers to be compared should be in the same column, not row. The tabulation of data follows the logic similar to figures: the independent variable should be on the left, and the dependent variable(s) on the right. Consider carefully, however, the organisation of the table itself. The table exists to present data;

arranging a table according to the leftmost column — a common practice — may not be the best way (see Box 12). A marginal indicator in the manuscript helps you to see if you mentioned all of the tables in the text, in appropriate sequence. In the title, one may indicate the organisational principle of the data on the table itself.

Tables are often wider than necessary. Consider if all the information needs its own column, or whether some can be combined as in the table in Box 12. This could reduce the width of the table, and may allow the table to be printed in portrait rather than landscape orientation.

Footnotes can give additional information to help understand the table. They should be short and to the point.

Box 12. A sample table

Table 1 Summary characteristics of the ground beetle assemblages along rural–urban urbanization gradients in nine northern temperate locations arranged according to their geographical latitude.

Country and urbanization stage	Number of		Trapping effort (trap-weeks)	Total number of		Activity density (ind./trap-week)	Number of forest		Relative frequency of			
	Traps	Weeks		Individuals	Species		Individuals	Species	Forest individuals	Forest species	Rare species*	Common species†
FINLAND‡	120	22	2640	2203	25	0.83	1520	14	0.69			
Rural	40		880	1167	21	1.33	695	13	0.60	0.62	0.57	0.24
Suburban	40		880	703	16	0.80	583	9	0.83	0.56	0.56	0.31
Urban	40		880	333	18	0.38	242	10	0.73	0.56	0.44	0.22
DENMARK	120	22	2640	10,319	43	3.91	5320	12	0.52			
Rural	40		880	4255	25	4.84	3151	11	0.74	0.44	0.64	0.16
Suburban	40		880	1670	25	1.90	1142	10	0.69	0.4	0.72	0.2
Urban	40		880	4394	37	4.99	1027	9	0.23	0.24	0.84	0.11
BELGIUM	78	26	2028	12,096	49	5.96	9490	22	0.78			
Rural	26		676	4047	36	5.99	3332	21	0.82	0.58	0.72	0.11
Suburban	26		676	3547	31	5.25	3026	18	0.85	0.58	0.77	0.13
Urban	26		676	4502	31	6.66	3132	15	0.70	0.48	0.74	0.1
ENGLAND	240	22	5280	10,648	36	2.02	10,600	20	1.00			
Rural	80		1760	2781	23	1.58	2772	16	0.99	0.7	0.65	0.17
Suburban	80		1760	4130	26	2.35	4106	17	0.99	0.65	0.81	0.12
Urban	80		1760	3737	24	2.12	3722	16	0.99	0.67	0.83	0.12
HUNGARY	120	34	4080	2140	50	0.52	1177	3	0.55			
Rural	40		1360	1206	25	0.89	867	3	0.72	0.12	0.6	0.08
Suburban	40		1360	457	26	0.34	246	3	0.54	0.12	0.54	0.19
Urban	40		1360	477	43	0.35	64	2	0.13	0.05	0.53	0.21
ROMANIA	120	22	2640	3651	38	1.38	2624	12	0.72			
Rural	40		880	999	19	1.14	929	11	0.93	0.58	0.47	0.21
Suburban	40		880	2352	22	2.67	1553	10	0.66	0.45	0.55	0.18
Urban	40		880	300	25	0.34	142	5	0.47	0.2	0.4	0.16
BULGARIA	132	24	3168	7035	72	2.22	5147	23	0.73			
Rural	24		1056	3125	45	2.96	2502	22	0.80	0.49	0.64	0.13
Suburban	24		1056	2210	36	2.09	1740	17	0.79	0.47	0.58	0.11
Urban	24		1056	1700	44	1.61	905	6	0.53	0.14	0.68	0.14
CANADA (total)	120	11	1320	15543	41	11.78	877	6	0.24			
Rural	40		440	1308	29	2.97	218	6	0.17	0.21	0.62	0.21
Suburban	40		440	3676	28	8.35	381	4	0.10	0.14	0.64	0.18
Urban	40		440	10,559	25	24.0	278	3	0.03	0.12	0.72	0.16
CANADA (natives)	120	11	1320	3628	37	2.75	877	6	0.24			
Rural	40		440	980	28	2.23	218	6	0.22	0.21	0.5	0.25
Suburban	40		440	1442	24	3.28	381	4	0.26	0.17	0.58	0.25
Urban	40		440	1206	21	2.74	278	3	0.23	0.14	0.57	0.24
JAPAN	120	22	2640	1627	26	0.62	1146	14	0.70			
Rural	40		880	882	23	1.00	670	12	0.76	0.52	0.57	0.17
Suburban	40		880	458	21	0.52	339	12	0.74	0.57	0.57	0.24
Urban	40		880	287	13	0.33	137	7	0.48	0.54	0.31	0.23

*Species with relative frequency < 0.01 were categorized as rare.
†Species with relative frequency > 0.05 were categorized as common.
‡Data from: Niemelä *et al.* (2002) (Finland, Bulgaria, Alberta, Canada); Elek & Lövei (2005) (Denmark); Gaublomme *et al.* (2005) (Belgium); Sadler *et al.* (2006) (England); Magura *et al.* (2004) (Hungary); Máthé & Balázs (2006) (Romania); Ishitani *et al.* (2003) (Japan).

Table 1 from Magura et al. (2010).

(continuous...)

This is Table 1 from Magura et al. (2010). Note that the information on the organisational logic of the table is included in the title. The column headings contain all information necessary to interpret the values in the body of the table, such as the measurement units. Where this is not possible or not easy (see the columns of rare and common species, for example, on the right of the table), footnotes are used. The two-level column headings allow some saving of space. On the leftmost column, the country and urbanisation stage were combined into one column, thus avoiding a repetitive column (writing out country names many times) and making the table narrower.

18

The First Version and the Process of Writing it

Most people can think, talk and write best in their mother tongue. However, writing your manuscript in your native language, and later translating it (or having it translated) is strongly discouraged (unless you want to publish it in that language). Believe the words of a one-time translator: translating a scientific manuscript is a very long and frustrating process. The article will have to be virtually re-written, because the idiomatic expressions are different from language to language. Moreover, scientific English — still the dominant language of scientific publishing — is not very complicated. My advice is clear and unequivocal: it is much better to improve a manuscript written in imperfect English than to translate an impeccable manuscript written in a language other than English. Always write in English; it helps if your notebooks are also in English.

The writing sequence will not usually follow the sequence in which the paper is structured. As explained previously, the writing process, for practical reasons, should start with the Material and Methods section. The last sections to be written will be the Summary/Abstract and the final title. It is also advisable that one works on the Results section together with figures and tables — as these should be tightly integrated. Very probably the Introduction will be written earlier than the Discussion; the latter can rarely be written before the Results are finished.

The Reference List should be built gradually. Do not leave this until the end, because this is a time-consuming approach that is prone to error. Whenever you want to cite a paper, immediately insert the relevant bibliographic data into the list. After all, good citation practice requires that you see the original article, to make sure you know that you have cited the paper's data correctly. It is very simple to add the bibliographic details at this point. Learn to use an electronic bibliographic database, as previously discussed. These can be linked to your text document, making the inclusion and listing of the citations much easier.

Do not forget that you can use paper and pen. The "loose-leaf-technique" is often useful: make separate folders for the different sections of the manuscript, and write key words, fragments, ideas, etc. on pieces of paper. These can be stored in the folders, and will serve you well when the "proper" writing begins.

Writing, even writing scientific articles, is a creative activity. From time to time, scientists suffer from the same proverbial syndromes as poets, writers, painters, etc.: the dreaded "block". You sit in front of your desk, or computer screen, and feel like nothing occurs to you and the paper will never be written. There are a few time-honoured tricks to overcome writer's block.

Start as soon as practicable: even with those half-sentences, ideas, sketches, fragments. They can be expanded later. When something is, seemingly, not progressing, do not get upset and insistent about it. Leave it for a while — after some time, hours or days, when you re-start, often the block has cleared, as if by itself.

Do not intend to write the "final" version of anything at the first sitting. For titles, for example, go through the suggested "provisional title—draft title—final title" sequence. You can do the same with other sections.

When you do not know where to start, you can get underway with insignificant details, such as acknowledgements, key words, addresses. These must be written and, even if they do not utilise your highest intellectual capacities, this activity is often enough to get you started.

Citing, Paraphrasing, Plagiarism and Self-Plagiarism

In the various parts of the paper, you will write about other people's ideas, results, theories. These should be acknowledged as such (i.e. the "intellectual property" of others) by citing the sources, in parentheses. The form of these citations has been discussed earlier. In the text describing these ideas, however, you should not, usually, use the discoverer's sentences — you have to express the same idea with your own words; you will *paraphrase*. If you find it necessary, you can *cite verbatim* — i.e. you present the idea as expressed by others, word-by-word. As the copyright of this text belongs to others, you must cite it with quote marks and the indication of the source, even if you only use a few sentences. The usual maximum that can be cited in this way without asking for specific permission is about three sentences — if the item is under traditional copyright. In a primary paper, it is customary not to cite *verbatim* even that much. Be careful because, if you do, you can be accused of *plagiarism.*

Plagiarism is when you use other people's work in your own, using the same words, and do not indicate the source. Thus, the text and the ideas seem to be yours, when they are not. This is ethically as well as legally unacceptable. It is little better than stealing — stealing other people's ideas and making them seem your own. The advice is clear and brief — don't do it. Your reputation will suffer irreparable damage. If a text is found to contain plagiarism, it renders the whole work invalid, and no journal will publish it. If it is published and plagiarism goes unnoticed, the repercussions are even more grave. The minimum consequence is that the journal will put the author team on its blacklist, never accepting any future manuscript from them for publication.

Today, several universities, as well as publishers, use various types of software to detect plagiarism. Perhaps it is a sign that plagiarism is more widespread than previously. Detected plagiarism carries a heavy penalty, and can cause great damage to you and your career. However, a further reason to avoid plagiarism is that it stunts your own intellectual growth. Instead of grappling with ideas and theories, understanding and expressing them in your own way, you would be making a shortcut that will prevent you from fully comprehending the ideas expressed. Remember — no two people use the same language when expressing the same idea. Strive for proper understanding — and this is demonstrated when you can write about the same idea, using your own words.

What about your own text, from earlier papers? You retain copyright — is it possible, legal, and appropriate to use this again? No, it is not — *self-plagiarism* is no better than plagiarism itself. The copyright of your own work may even have been signed away to a publisher. In any case, originality is important in science — and repeating yourself, word-for-word, is not allowed; you should avoid self-plagiarism. This may not seem sensible advice but, if you feel you have to use the same text in the introduction of three or more of your papers, you are probably trying to publish the least publishable parts — and will frequently fail to reach the publishing threshold.

Above all, be careful: the joke that "Stealing from one is plagiarism, stealing from many is research" is, emphatically, *not* true.

Completed? Not Finished

When you complete the writing, you are not yet finished. Never submit the first, freshly-completed version of your manuscript. The chances are that the manuscript contains some of the common errors listed in Box 13. First: read it yourself, from the beginning to the end. You have spent a lot of time on different parts, working on them in odd sequences, leaving and returning. Be your own first reader. Does it read fluently? Can you spot errors, omissions, inconsistencies? Are all parts complete? Have you referred to all figures and tables, and are they numbered in the sequence in which they occur? Are all references mentioned in the text on the reference list and vice versa?

Box 13. The most common mistakes in a newly completed manuscript

1. Haste

When the manuscript is completed, the author usually feels almost fed up with it, and wants to see it off her desk, submitted immediately. Premature submission is one common mistake. Solution: wait. Send the manuscript to one or two more colleagues and wait until they return the manuscript with their comments. Never send the manuscript until all such comments came back (or the colleague told you she cannot comment on it) and you carefully evaluated them; this will usually result in revising your manuscript. The end results will be an improved manuscript.

2. Confusing finish

Also towards the end of the writing process, the clarity of argument decreases; the argument is getting muddled. This is a sign of tiredness. This is when results sneak into Discussion, methods into Results, and non-written result statements into summaries. I think this also arises because the authors "see the light at the end of the tunnel" and this makes them rush. Solution: make a little graphical representation of the flow of your argument to help you to evaluate its effectiveness and clarity. Does one point indeed follow from the other one? Is the structure consistent?

3 Inconsistent sequence/structure

This results from the (otherwise natural) writing of different bits at different times. Experiments should be described, and their results presented and discussed in the same sequence, even though the different parts (hypothesis, methods, results, significance) belong to different sections of the paper. However, details of experiment 1 should always precede those of experiment 2 in all the major parts. Solution: again, make a little diagram and check: are all parts of all experiments presented in the respective parts of the manuscript? Is the sequence consistent?

4. Not enough detail

This occurs mainly in the methods. Omission of elementary information — because it is common in the lab, and is a basic method in the field. Nevertheless, it still needs to be written in detail (or referenced). Solution: give the Ms to a colleague and ask if she could repeat the experiments on the basis of your description? This often brings out omissions that you can then correct.

Important detail may also be missing because the writer has no clear concept of her future reader, and thus information that would be helpful is not presented. Solution: always write for a precisely identified journal, and familiarise yourself with the circle of readership. Give enough information for her to understand your new results.

5. Lack of clarity concerning in-text quotations/references:

This includes writing WHO did something and not WHAT was done. Evaluating other people's work (XX's brilliant experiments, ZZ's pathbreaking approach, etc.). Solution: always write WHY a publication is cited, but restrict the statement to facts.

6. Data-poor, badly designed figures

The figure is there because "a proper paper has figures". Solution: critically revise figures, first asking: is the figure necessary? The most important further question should be when revising figures: can the data : ink ratio be improved?

7. Errors in references

Omission of a reference, no total overlap between cited references and the list, incomplete references with parts (initials, volume or page numbers) missing.

Solution: learn to use a literature database (Reference Manager, EndNote, Zotero, etc.), and use it in conjunction of your writing program. Generating the reference list can then be automated, dramatically reducing the number of errors in the manuscript.

8. Formatting errors/mistakes in the manuscript.

This also occurs due to haste to complete and submit. Needless to say this hinders, not speeds up publication: the first check on a new manuscript is on its format, and if this is not as required, the manuscript will be returned without evaluation.

When your manuscript has passed your own first test, it is a good idea to show it to others before submission. This "pre-submission peer review" could include three kinds of readers. Show it to a friend, who is at least somewhat familiar with the area. We rarely cherish criticism, and we often take this more readily if coming from a friend. If possible, also show it to a colleague. If you write for a more general readership, it is wise to show it to someone who is not closely familiar with the area, maybe from another profession.

There are no rules here — you can show your manuscript to as many people as you want. This counts for nothing at the journal when you submit. Nevertheless, the manuscript will benefit from such independent, and usually benevolent, advice. You can also send it to a colleague, whom you have not met before. It is polite to ask first, though.

Exposing your manuscript to such "unofficial review" is also a good way to improve your writing skills — especially if you have a patient, experienced colleague who can go through your manuscript and explain the points she criticised.

Collect the comments and revise your manuscript. You do not have to accept any advice — the work will be published under your name, after all. However, most of the advice will be given with an intent to help and, very likely, at least some of the assistance will be worth accepting. When this revision is done, put the manuscript aside to "mature". Authors are, naturally, very impatient at this stage. However, I suggest that you do not rush; a few days' rest can mature the paper.

PART III

The Scientific Paper and the Process of its Publishing

19

How to Prepare the Final Version

"When I have finished the experiment, and written the paper, the final formatting is not important, because if my work is good, solid science, it will be accepted for publication". If you believe this statement is correct, you are *very* wrong.

Why? This can be understood from the working practices of today's scientific journals. Scientific editing in most journals is still done by volunteer scientists. They are not keen on spending a lot of time on formatting. Moreover, all journals have a specific format, and all papers printed must follow that format. Consequently, the format of a submitted manuscript is among very first things to be checked. If it does not fit the format required by the journal, the manuscript will be sent back *without evaluation,* and the author is asked to correct the format to fit the requirements. Be careful in this case — the authors can only fail once. If the author does not correct the format properly, the editor will probably not only send it back, but may blacklist the author team and advise them to direct future publishing attempts elsewhere.

This attitude also conforms to the requirements of optimal decoding — if the coding (i.e. the format, in this case) of all the papers is the same, the readers will be able to digest the content faster. They do not have to waste time looking for certain information (methods, question, or conclusions), because they can always find it in a familiar place, which makes understanding faster and easier.

Manuscript formatting remains largely unchanged even after the rapid spread of electronic submission. A manuscript should be double spaced, with wide margins. No right justification is needed. As for formatting, no mimicking of the final journal format is needed. Once the manuscript is accepted, the copy editor will make sure the format corresponds to the journal requirements. At this point in the process, mimicking the final format would be an unnecessary diversion, and a waste of time on the author's part. Only the following elements of the formatting should be observed (also, check the relevant Instructions to Authors as variations are possible):

- If a text line is centred, centre it in the manuscript.
- If text appears in bold in the journal, it should be the same in a manuscript.
- Indicate Italicised text by either underlining, or setting the relevant text in Italics
- All characters that will be in capitals, or small capitals, in the published paper should be written in that style in the manuscript
- Observe the heading structure and follow it in the manuscript.
- The journal reference style must be followed.

Further, it is a good idea to:

- Paginate the manuscript, with the author's name at top right (or "first author-et-al.", if appropriate), as a header.
- Start new sections on a new page.
- Check the spelling. The two major spellings of English, the British and the American, are not the same. They seem to be mutually irritated by the other's spelling practice (remember the saying: the British and the Americans are divided by a common language). Be careful as no cutting of corners is allowed here — an American journal will insist on American spelling, and a UK journal will require the British equivalent.
- Use numbered lines in the manuscript — this makes it easy to follow comments and revisions for the editor, reviewer(s) and author. The line numbering should be set to continuous — beware, because the default in Word is "restart on every page".

You must also place tables and figures at the end; do not insert them into the text, unless the journal instructions specifically ask for this. The position of tables and figures will be decided by the technical editor, when typesetting the text. You should, however, indicate the approximate, desired position, either by inserting a box, or using a note in the margin.

Should you deposit your MS in an open archive?

Once your manuscript is complete, you can decide to send it to a prepublication manuscript archive. There are several of those, and they are usually field-specific. In biology, for example, BioRxiv is the oldest and biggest. Some of them allow others to attach comments but as a rule, they are not there for formal or informal review. The advantage is that such manuscripts obtain a DOI, and thus become citable. In case of a later dispute, this can also serve to decide about priority. They have not been peer-reviewed, so they do not count as "valid" publications. Call me a conservative, but I prefer my own work to benefit from the critical assessment of my peers before I make it fully public. But the choice is, of course, yours when it comes to *your* unpublished manuscripts.

20

The Manuscript and its Submission

Once you believe the manuscript is ready to be sent, it is still worth being careful. Now, your natural desire is to see your work published as soon as possible; however, from this point, the manuscript will have to be handled by several people before publication and, therefore, extra care on your side will speed things up.

Electronic Submission

Many journals, and most of the journals published in developed countries, require that you submit your manuscript electronically. This takes time — sometimes a lot of time. Prepare your manuscript carefully. Several journals provide you with a checklist (see Box 14 for an example) — it is a good idea to save one of those and use it, even when the target journal does not have one. Once all the parts are together, you should first register as a user on the journal website. Do this with thorough attention, checking all the questions or options, as well as what you authorise the website to use your data for.

Box 14. Pre-submission checklist

1. Is the manuscript complete? Does it contain all the necessary parts, arranged in the necessary order?

2. Are all word limits observed? There can be limits by number of words or even number of spaces on the total article, title words, abstract, highlights, key words, and the running title.

3. Is the summary well structured, informative?

4. Are the necessary structural requirements observed? These can include primary-secondary-tertiary headings, placements, numbering, etc.

5. Does the numbering of figures and tables correspond to the sequence of being mentioned in the text? Did you indicate the desired position of all figures and tables?

6. Are all listed references cited in text and all that are cited in text have the full bibliographic reference listed?

7. Are bibliographic details correct and in the appropriate format?

Once you have registered and chosen a password, you should log in, and start the submission process. The process is step-by-step, and largely self-explanatory. Usually, all steps are obligatory. During electronic submission, the submitting author must do most of the clerical work that used to be done by the journal personnel. The submission process is tedious, often non-intuitive, and always takes a lot of time. Additionally, it depends on the speed of your Internet connection. The publisher wants you to believe that this system will allow you to track the path of your manuscript — but it is not actually much help. The editorial team will not want to hear from impatient authors immediately post-submission: "Leave us in peace, we are working on your manuscript". The author, preferably, would like a daily update on her manuscript. Here the editor's interest prevails — for months, all you may see is that your manuscript is "in review".

When submitting (unless the journal requires otherwise), combine your manuscript into one file. The figures do not have to be in their finest resolution, especially if this necessitates the use of different programs and file types — it is much easier if everything is together. You can attach figures to a Word document and, if possible, do so. When naming the file, include the corresponding author's name.

During the submission process, you should normally enter all the authors with their affiliations, the title, abstract, keywords, length of your manuscript, suggested and non-preferred reviewers, and various other bits of information that are sometimes not easy to link with your actual manuscript. Finally, you upload the manuscript itself, which will be converted to a pdf file. You will have to open and approve this. Only then will the manuscript be finally submitted. It is a good idea to save a copy of this pdf file.

Do not forget about the accompanying letter. This can sometimes be added as a separate file — but, even if not, there will be space to send a message to the editor.

The Accompanying Letter

Never send anything without an accompanying letter. This is not only a matter of courtesy; it is in your, the author's, best interest to attach a cover letter, as it can speed up the handling of your manuscript. Editorial offices often resemble organised mayhem; they may handle several journals, and sometimes receive hundreds of manuscripts per day. Some are first submissions, others are annotated manuscripts, sent back by reviewers, still others are revised versions, or final manuscripts, sent by authors. The accompanying letter helps the office to handle your manuscript more effectively — and this means processing it more speedily.

In the accompanying letter (see a sample in Box 15), state the following:

- What is being submitted? Is your work a new submission, a revised manuscript, or a final version?
- For which journal? Offices may run many journals and, from the title, it may not be obvious which one your work has been submitted for. You cannot ask the office to decide where your manuscript could fit; you have to make that decision (see Chapter 4).
- Declare if any part of the work has previously been published and, if so, which parts and where? If it was presented as a talk, or a poster at a conference, it is also worth mentioning. An abstract in a conference volume is *not* a publication, but may be worth a mention. Be careful with Internet publishing — while many journals allow pre-publication in an open manuscript depository (such as BioRxiv: https://www.biorxiv.org/), others even if you put the manuscript on your personal website, consider it published, and will not touch it.

- Indicate the name and address of the corresponding author. Even if this is already noted on the manuscript, redundancy here is acceptable, even welcome. The editor does not then have to look up the address from the manuscript. Also, indicate if the corresponding author will have a different address during the next 6-8 months, even if temporarily. This will help the editor to get in touch with the corresponding author without delay. Do not go overboard — there is no need to let the editor know when, and where, you are going on holiday for 2 weeks. However, if you will be away from your workplace for more than a month, it is worth letting the editorial office know this, and give the temporary address. Even if they cannot reach you, at least they will know the reason for your lack of reply.
- You should also state the co-authors' agreement. This declaration will have to be repeated when the paper is accepted and the copyright form is signed. Nevertheless, it is also required here.
- State the uniqueness of the work, indicating that this manuscript contains new, unpublished results.
- State that the work is not under consideration elsewhere. You cannot send the same work to more than one journal at any one time.
- Finally, in one or two paragraphs, argue for the merits of the manuscript. Journals have become so overloaded that the first decision is often made by the editor, who, after a quick scan of the paper, decides whether the manuscript should go out to reviewers, or be rejected without review. Assist the editor by pointing out the major new findings in your manuscript, and provide reasons why it should be considered for detailed review. This summary should not contain sentences "cut and pasted" from your manuscript — rephrase them.

Box 15. A sample manuscript submission letter

The Editor

Global Ecology & Biogeography

17 May 2008

Dear Editor

Re.: Submission of a MS

Enclosed please find a manuscript by Magura et al. "Urbanisation decreases the diversity of forest specialists but not overall diversity in ground beetles (Carabidae)". We would be grateful if this could be considered for publication in the journal "Global Ecology and Biogeography".

This work reports results obtained in the international Globenet Project. The Globenet Project, started in 2000, is an attempt to find out whether urbanisation has a general impact on invertebrates in geographically different locations. The individual sub-projects have a common setup and methodology, creating the necessary pre-conditions for a synthesis that has not yet been done. In this manuscript we attempted to examine one general hypothesis: that urbanisation would be detrimental to biodiversity. We believe the specific new aspects of our work are the following:

1.) We analysed patterns of diversity by using diversity ordering by the Renyi diversity index, which is not commonly used in ecology, although it has recently been recommended by Southwood & Henderson (Ecological methods, 2000) as a preferred way to compare diversity trends. We demonstrated that even by this synthetic measure, diversity does not decrease as urbanisation intensifies.

2.) As a new approach, we separately analysed the forest-associated species. If only the sub-assemblage containing species linked to the original habitat (forest) is considered, the diversity ordering indicated a decrease along the urbanisation gradient.

3.) Finally, we found that while urbanisation has a documented homogenising influence on the flora and fauna of cities, this does not seem to happen in ground beetles.

We believe that the results briefly summarised above reveal so far hidden effects of urbanisation on carabid assemblages, contradict some accepted wisdom, and represent an advance in the understanding of the effect of urbanisation on biodiversity. We believe that Global Ecology & Biogeography is a suitable forum for this research.

This is a first submission, containing unpublished information, and is not under consideration elsewhere. All co-authors have read and agreed to the content of the manuscript.

Please address all correspondence to me at the address indicated on the manuscript. Thank you in advance for your editorial assistance.

Sincerely yours

XX

Submitting author

Address

Submitting by Mail

Today, electronic submission is the usual practice, although a few journals still operate "on paper". To these journals, you have to submit your manuscript by mail. Your manuscript will be ten or more pages, and you are routinely requested to send three or four copies. This makes it a rather bulky shipment. If it is in a flimsy envelope, this may tear; parts of the manuscript can get damaged, arrive in battered condition, or lost; all this may slow the handling and publication process. If you do have to submit paper copies of your manuscript, use a strong envelope, possibly a padded one, and make sure the edges are strong enough or protected by extra tape. You can even use duct tape to strengthen the sides and corners — this is where a bulky envelope will be most easily damaged. Carefully separate the different copies — but do not staple unless the journal specifically requires you to do so.

Send the manuscript by air mail (where required) and pay the appropriate postage charges. Mail services often downgrade mail with inappropriate postage and your manuscript may spend weeks or months, instead of days, in transport.

Keep a hard copy for yourself. This will be your insurance against computer crashes and other unforeseen complications. This will also be the proof of the existence of the work — in many cases, electronic copies are not accepted as proof.

Check that you have the required number of *full* copies. It is probable that different parts of your manuscript (for example, the text and figures) will be created using different programs and, thus, it is not always sufficient to print X copies of the same document. Be meticulous. Many journals provide help by offering a pre-submission check-sheet — use these.

21

Handling a Manuscript and its Associated Procedures

The manuscripts are handled in a systematic way by the editorial office. This process is usually lengthy and involves several people, with repeated correspondence between the corresponding author and the editorial office. It is important to know, however, that correspondence is always between the editor and the corresponding author, and that the editor will not contact anyone else from the author team. It is even more imperative that, if you are a corresponding author, you correspond *only* with the editor or handling editor, and no one else. Everything goes through the editorial office: revisions, enquiries, complaints, etc. Breaking these rules will result in the immediate rejection of your manuscript. Do not risk it.

The manuscript will make its way to publication through the following steps:

1. When the manuscript is received, the first task is for the publisher to register it in their editorial database. This administrative process will include entering the arrival date, authors' names, corresponding author and their address; the manuscript will then receive a reference number. This reference number is important as this is the identification tag for the manuscript during its further processing. When submitting electronically, authors must perform these tasks themselves.

2. After this, the manuscript is briefly checked: does it fit the theoretical scope of the journal? This leads to the first decision. If the topic does not fit the journal, it will be immediately sent back to the author without further consideration. In theory, the authors can appeal against this

decision, but it is not usually worth it. Editors typically have a very firm perception of what types of papers they want in "their" journal.

3. Next, the format of the manuscript is checked. Does it conform to the required format? Is the text appropriately structured? Are word limits observed? Is the recommended position of the tables and figures indicated? This leads to decision no.2: manuscripts that deviate grossly from the expected format are also sent back with a request for correction. This should never happen to a carefully prepared manuscript.

4. This is followed by a quick reading of the manuscript, usually by the editor, to decide whether, or not, the manuscript should be sent for detailed review. Many submitted manuscripts are rejected at this stage. The editor's decision is often based on non-scientific reasons, editorial policy, etc. — it rarely rests on detailed scientific scrutiny.

5. If the manuscript passes this stage, the editor usually assigns it to a sub-editor (or handling editor). The sub-editor's (or handling editor) task is to find suitable reviewers, send the manuscript for review, and keep track of its progress. Once reviewers have been found, the handling editor sends them the manuscript, indicates a deadline for the review, and receives (or, if necessary, solicits) the reviews.

6. Once the written reviews are received, the handling editor evaluates them, and makes a recommendation to the editor (or decides herself) whether the manuscript should be accepted, rejected, or revised. In cases of seriously divergent reviewer comments, additional reviewers may be asked to give their opinion on the manuscript.

7.1. If the decision is to reject, this must be accompanied by the written reviews. It is the editor's right to make final decisions about acceptance or rejection but she is also obliged to justify the decision. Never accept a simple decision without justification.

7.2. If the decision is immediate acceptance (which is rare), the editor will ask you to submit the final version. You should do this promptly.

7.3. Usually, even accepted manuscripts will have to be revised. The authors are typically given a deadline to submit their revised manuscript. Once the revision is returned to the editorial office (or re-uploaded on the website), the handling editor will evaluate the revision, sometimes sending it out again to reviewers.

8. The handling editor now decides more quickly about acceptance, rejection (yes, it still is possible), or indicates a need for further revision. There may be further rounds of correspondence between editor and the authors at this stage, not usually involving the reviewers further.

9. Once the manuscript is accepted, the handling editor will normally send everything to the editor, who manages the final version. She will notify the authors of the acceptance, invite them to submit the final version of the paper and, when received, check the format, language, figures, and send the now-accepted manuscript to the printer.

Contact with the Editor During the Review Period

Authors are not normally expected to engage in follow-up correspondence after submission. They should be aware, however, of how the process runs, and there *may* be a reason to contact the editorial office. Firstly, the author should hear about the outcome of the first two screening processes: covering the topic fit and the format evaluation. Many journals do not send out a notification before the manuscript also passes the third screening, and is sent to reviewers. This should not take more than one month, so if you do not hear about your submission by then, it is acceptable to contact the editorial office. Something may have gone wrong.

When the submitted manuscript is sent to reviewers, the editor usually indicates a tentative deadline for review. Note that this deadline is only tentative, as most individuals involved in the review process work as unpaid volunteers and, thus, the editor has limited powers to convince them to keep to these deadlines. The period of review can vary between 6 weeks (which only the fastest journals can keep) to 5 months.

If you do not hear about your manuscript ca. 4 weeks after the indicated review deadline, it is acceptable to send a polite message of enquiry. You can unearth a problem, prompt an editorial action, or get clarification as to which stage the manuscript has reached in the process.

If, after repeated attempts, you do not receive a reply, remember that until you sign a copyright transfer form, the manuscript belongs to you, the authors. You can withdraw the manuscript, which means the journal in question does not have the right to publish it. If you decide to do this, do it in writing, sending a registered letter. It is sufficient to write that you intend to withdraw the manuscript — you do not have to give a reason. Try to be polite, even in this case. Withdrawal is also advised if an error, or worse, suspected data falsification is found in the work; this is better than having to publish a retraction or correction.

22

The Editor's Report and its Reception

One day, you will receive an editor's letter with the decision. If your manuscript is accepted, send in the final version promptly, then celebrate. You will not have many such occasions during your career.

Not infrequently, you will receive a rejection letter. When such a letter arrives, it is never the brightest of days. However, do not do anything hasty, and do *not* lash out at anyone. After feeling sorry for yourself, start thinking and acting. Firstly, remember that you are in ample company. No one really knows the precise figure, but anecdotal evidence holds that an estimated 66% of published papers are not published in the journal where they were first submitted as manuscripts. Much publishable work is rejected by journals; the rejection of your manuscript does not necessarily mean that it is unpublishable.

Secondly, do not take any criticism personally — maybe a piece of your work was not up to a required standard, but this is no judgement about yourself as a scientist, nor as a person. In fact, good reviews make this very clear.

Next, re-read the editor's and reviewers' letters carefully (insist on receiving the reviewers' reports if not included. Avoid journals that do not send them to you, even after prompting). These can contain valuable suggestions. If the reviews are not appropriately phrased, remember that the impolite reviewer was still using his/her spare time to provide comments, and s/he may well believe that s/he is being helpful. Look beyond the form and concentrate on the content.

There are different types of rejection letters. The first can be called the "never want to see it again" type. The editor's letter makes it very clear that the manuscript is not acceptable. It is rarely worth arguing, even if you feel this judgement is incorrect. It may be possible to convince the editor that the review was unfair but, in most cases, it is simply not worth the time and effort. Send the manuscript elsewhere (after review and the necessary format modifications).

However, the rejection may not be this final. After completing any additional experiments suggested, the editor may be willing to consider the work again, but as a new submission. This may or may not be indicated in the decision letter. You have to think very carefully about whether the required work is worth it; it may still be a better option to send the manuscript elsewhere. If you believe your manuscript might be given a second chance (after revision), enquire from the editor whether she would agree to a re-submission. Some journals have a policy that a rejected manuscript cannot be resubmitted. In that case, there is no room for negotiation.

Thirdly, the letter rejects *the current version* but invites you to revise and re-submit. This is often equivalent to a major revision. The wording is important; today, many editors are unwilling to indicate that the work is basically publishable, and only write that it needs additional work.

If your work is rejected on what you believe, or can even prove, are insufficient grounds, it is, again, best not to argue. A good editor knows that her reviewers are not infallible, yet the process is very advanced, and is not easily reversible; the chances are that your paper will not be best placed in that journal, anyway.

Several of the most important journals receive a huge influx of manuscripts, and the first, quick decision is taken by the editor whether or not to send the manuscript out for peer review. Most manuscripts that are received are not sent to reviewers, but rejected after a short evaluation by a single person. This, while it increases the chances of misjudgement, seems necessary due to the sheer number of manuscripts involved. If your work is rejected via this mechanism, it is not worth arguing.

23

Writing Revisions to the Papers

At this point, let us think about editors. First, jokingly (I am paraphrasing Day's (1999) joke here): "The editor dies and goes to heaven. After due scrutiny of his deeds and some questions concerning his profession from a puzzled St. Peter, he is allocated a small apartment with a balcony. Here he adopts the habit of enjoying the sun that always shines in Paradise. One afternoon, a pope walks nearby, and, spotting him, he asks who he is. More questions follow, as the puzzled pope obviously has no idea what an editor does. The conversation does not last long, though; the pope curtly says farewell, and proceeds to St. Peter. There, he starts complaining about his own accommodation, which is smaller and without a balcony. "This is not fair" — says he — "after all, I did my earthly duties well, on *your* behalf; that is why I am here. And then... this lowly editor... gets such lavish treatment"... to which St. Peter replies: "Well, acknowledging your good deeds, there is nothing so overwhelmingly special about you. There are about fifty popes here. But this is the *first* editor who made it to heaven!".

The editor and her crew of reviewers are sometimes viewed by aspiring authors as the defenders of that desirable castle, the journal. They are "in" and their task is to keep as many of the aspiring authors as possible "out". Get rid of this combative attitude. Allow more editors to reach heaven — do not send them to the opposite side of eternity. The interests of the editor and that of the authors are the same: to publish the highest quality new results, in the most understandable and convincing way. If that were not so, you would soon find yourself like the American actor Groucho Marx, who is claimed to have stated: "I do not want to belong to a club that accepts me as a member". You will not want to publish in a journal that accepts everything you write without hesitation or correction.

Remember though, when the outcome of the review is a request for revision, this is not an obligation. Check the suggestions, requests or recommendations carefully. Is the manuscript worth revising? The answer is: not always. Remember too, that until you sign a copyright transfer form, you can decide about the fate of the manuscript — including withdrawal from a journal.

Once you decide that you will revise a manuscript, avoid two extreme attitudes:

The first is trying to prove, however systematically, carefully or methodically, that every single comment, criticism or recommendation is wrong, and you do not want to accept any of them. This is the "zero acceptance" standpoint. An editor will not accept this approach.

Do not believe, either, that if you accept all suggestions without dispute (the "100% acceptance" attitude), the editor will be happy, and the manuscript will swiftly be accepted. Such a slavish attitude will not get your work published more easily, especially not in good journals.

An editor expects careful consideration of the reviewers' suggestions so that a compromise emerges, and she is quite open to partial acceptance of the reviewer suggestions. In her letter of evaluation, she often indicates which of the suggestions should be considered carefully. The editor appreciates the work of her reviewers, but knows that they are not always correct. Thus, you are not obliged to accept all suggestions. In some cases, they are contradictory anyway, so you will have to decide — you cannot accept both. An additional fact is that you, the author, are probably pretty much convinced that the manuscript is well written, and naturally reluctant to re-write it extensively.

Give careful consideration to the reviews, especially if they are good and factual. Sound criticism contains three elements. It indicates what is, in the opinion of the reviewer, incorrect or wrong. Second, it supports this opinion with arguments. Third, it indicates how to improve the areas of concern. If a reviewer only indicates her dislike, or disagreement, about something you wrote, but provides no justification, I suggest you acknowledge the difference, but do not change anything — the reviewer must justify her criticism. Simply declaring that your opinion happens to be different, or that you disagree with something the authors wrote, does not advance science.

It is natural that the author feels that the phrasing, tests chosen, and other, often criticised, elements are well phrased, calculated, or carried out. However, do not be rigid about this. While it is natural that you want to avoid extensive re-writing, accept suggestions if you feel they are not inferior to your original ones.

You do not have to agree to every suggestion (and I repeat: this will not help you to get your manuscript accepted); be careful and objective when you do not want to accept a suggested change. The best course of argument is to cite published evidence in support of your stand- or viewpoint. In such cases, the reviewer's opinion is countered by published evidence. This implies not only that another author shared your view or came to the same conclusion — so did the reviewers and editor(s) of that published paper. Thus, there is a group to support your view, method choice, or result. Remember that different types of published information carry different weight; try to find the highest quality evidence to support your view — this would be a peer-reviewed primary paper.

The key to a good revision lies in considering the editor's work. How does she treat your revised manuscript? Once it arrives, potentially, she has to scan and compare several documents: the original manuscript, the original comments, your replies and the revised manuscript. One by one, every single one of them. This is complicated work, and a task that requires concentration and time. You can help the editor if you prepare a very detailed reply that combines the original reviewer comments and your replies, point-by-point. Reply to every single comment. If relevant, do not only write your reaction, but indicate the place or line number where you changed your manuscript and how. Naturally, indicate the line number of the relevant sentence of the revised version. If you changed a sentence, it may be a good idea to cut-and-paste it from the revised manuscript. This way, the editor has most of the necessary information collected into one document, and this makes her work easier. Also, you may be remembered fondly as the rare author who considers what others must do to get her work into print. You cannot bribe the editor — but you can gain her goodwill, and that is not to be neglected.

Above all, remember that the peer review process is voluntary, and that all those people often used their own free time to try to help you to improve the presentation of your work. Reviewers and editors are often highly experienced, intelligent people, with more than a passing understanding of the topic of your work. Their comments will nearly always improve your paper.

The process is long, not without frustration, and is far from perfect. Yet the editor is on your side, and strives to get the best quality from every manuscript that is submitted. If you really feel you must object, be as restrained as you possibly can. If you must, blame the reviewer, not the editor.

24

The Final Version of the Paper and How to Submit it

When the content is finally found to be acceptable, the editor will send you the much-awaited message that your manuscript is now accepted for publication. She will ask you to submit the final version. I recommend that this is done before the celebration begins, and there are a few important things to remember:

No Changes

Firstly, now you have reached an agreement with the editor about the content of your paper, there is absolutely no freedom to change *anything*. Even if you discover a printing mistake that escaped everyone's attention so far, you must point this out to the editor when submitting the final version. Be very conscientious about this.

Be Prompt

If you check the acceptance dates of published papers, you will realise that the publication sequence does not correspond to the acceptance dates. After acceptance, papers are handled simply on a "first come, first served" basis. Also, remember that papers are frequently not published continuously, but as a group, forming an issue. When an issue is full, all accepted papers that arrive subsequently are pushed into the next issue(s). The publication queue can be quite long; I have seen delays of up to 8 months. Your promptness may significantly affect when your paper is published.

Check any detail that may have changed during the evaluation process; update them. This means, most frequently, altering the citation details for papers that were cited as "in press" and are now published.

The current address of some authors may also have changed; these need to be updated.

Send Final Copies of Figures

Now, in (the very rare) case that you have hard copies of photos or drawings, send the best-quality version at this stage; package them carefully so that they do not get damaged in the mail. You will probably have to submit electronic versions of your figures, usually in separate files, and in prescribed quality/resolution or file format. Follow the journal guidelines carefully (most publishers prefer vector-based graphical material), and always try them out on your own computer to make sure that you are satisfied with them. This, however, does not absolve you from checking the proofs very carefully (see later).

Data Access

There is an increasing trend that authors are requested to make their raw data accessible to others. This can be done by depositing your data in an open access depository (which is often supported by the journal), and/or submitting them to one of the increasing number of data journals. In many journals, access to data is a precondition for publication. Currently, authors are allowed to declare that they will provide the raw data on request, but I believe this is a temporary arrangement, and open data access will soon become the norm.

Sign Any Necessary Forms

At this stage, if not earlier, you will have to sign a copyright transfer form, probably electronically. This is necessary; without this, the publisher is not at liberty to publish the paper — you, or your team, are still the legal copyright holders. The copyright transfer form is usually a standard one, in which you agree not only to transfer the copyright to the publisher, but make a legally binding declaration of other key facts: that your team agrees with the content, that the content is new, and has not previously been published. You should read this form carefully and make sure you understand what you are agreeing to.

This step will not be necessary if the journal is Open Access and uses Creative Commons licences. In this case, authors retain the copyright to their work, and anyone can freely use, cite, copy or distribute it according to the terms of the licence, providing the source is clearly acknowledged.

Make Sure Your Documents Are Readable

When submitting the final version of your paper, always check that no file is corrupted, they can all be opened, and that you have used meaningful file names. It is a good idea to indicate, in the accompanying letter, the names of the files and the program (including version number) that was used to generate the files.

Do Not Forget the Accompanying Letter

Do not send anything to the editorial office without an accompanying letter — even at this juncture. The letter should indicate the journal name, and that it is the final version that you are submitting. Include the title, authors, corresponding author's address, again, even if it may seem redundant — consider this as an extra assurance that the proofs will be sent to the correct address. State the names of the file(s), the program used to produce them, and the version. This is especially important for graphical files. If there are any additions or errors discovered and corrected, state them and indicate their position in the manuscript. You may decide to modify the acknowledgements, for example, mentioning the reviewers if you found that their comments improved the manuscript. It is a polite gesture to thank the handling editor for her work, even if informally, in the letter — this is all too infrequent. Remember — you have been helped by many people, not machines. They may have done their (paid) work, but almost certainly, several of them were volunteers, using their own time. The world is not as large as it seems to you, especially when you start your scientific career. The whole process of publication relies upon teamwork, and team members will remember you and your general attitude. To be polite is not mere courtesy, but also a smart investment for the future. For an example of a letter accompanying the final submission, see Box 16.

Box 16. A sample accompanying letter to final submission

Dr. XX XXX

Editor, Journal of Unrepeatable Studies

Department of Probability

The Unseen University

XX Town

23 July 2013

Dear Editor

Re. final submission of the MS Nr. JUS-2012-1122

Enclosed please the final version of the paper "XX.. XX" by AA, BB and CC, accepted for publication in the JUS on XX Dec 2014. We uploaded all files as instructed by the editorial program. Our final submission contains three files:

1. "Careful-JUS-2012-1122.doc" is the text file, containing all parts of text. This was written using Word version 2010

2. "Careful-fig1-JUS-2012-1122.tif" is an electronic version of our figure 1, prepared using Adobe Acrobat version 13.0

3. "Careful-tabl1-JUS-2012-1122.tif" is a copy of our table 1, also generated using Adobe Acrobat v.13.0

We discovered two spelling mistakes that we corrected (lines 233, 255) — these are the only changes with respect to the earlier, accepted version.

Please send proofs to me, the corresponding author, at: Department of Miracles, Underwater University, Coolabana, West Indies 22009.

Finally, thank you for your editorial assistance to help our paper to publication. We remain available as eventual reviewers for JUS.

Sincerely yours

XXX

Corresponding author

25

Processes after the Acceptance of the Manuscript

The Technical Editing and Printing Process

Once the scientific editor receives the final version, she carefully checks the final version vs. the accepted manuscript. If she is satisfied, the manuscript is forwarded to the printer. The accepted manuscript first goes to the copy-editor, who checks the spelling and punctuation, the style of the journal, unifies the abbreviations, and "marks up the manuscript". This is a process of inserting comments and instruction on the margin of the manuscript, indicating the typeface, size, position, arrangement of figures and tables. These instructions are for the printer who will do the typesetting. If something is unclear, she will insert so-called "author queries" — points of clarification requested from the author.

Based on these instructions, the printer typesets the manuscript and inserts the figures and tables, producing the first proofs (in American English: "galley proofs"). These proofs look like the final, published article, with all the letter types, sizes, and arrangements as they will appear once the article is published. A few details may be missing; for example, the final page numbers. There are usually also line numbers that will not be printed in the final publication.

These proofs (with the author queries if relevant) are sent to the corresponding author for proofreading. Once the corrected proofs are returned, the printer performs any necessary corrections and produces the second proofs. These are also checked, but usually only by the production editor, not by the author. These second proofs also go back to the printer, and any remaining final corrections are made.

After this, the whole issue is collated, the final page numbers inserted, and all of this is sent for final authorisation to the editor, who, after checking one final time, authorises printing or publication. The journal is printed, bound and mailed (if published on paper), or put on the journal website. The publication process is complete; your article is published.

Proofreading

If you believe that, after sending a perfect manuscript as final submission, a perfect paper will result, because all further work is done by computers, you are *very* wrong! In my long publishing career, I have not published a single paper where there was no correction to be made at the proof stage. Your opportunity to check the appearance and quality of your paper is when the publisher sends you the proofs. This is your last and only chance, so you should do the proofreading very carefully and attentively.

Authors will receive a very short deadline for proofreading, usually not more than 72 hours. Publishers are very keen to publish as quickly as possible and therefore such short turnaround times are routine. You should receive a copy of your final submission, which formed the basis of the proofs, and the proofs themselves (two copies, if on paper). These may be accompanied by a so-called author query sheet. You are asked to check if the product (the proofs) matches that of the original (the final manuscript), and indicate any errors or mistakes that have occurred. With the rapid spread of electronic proofreading systems, all you may get is the proofs — one more reason to keep a copy of your final submission.

How to Do the Proofreading

First, do not rely on a spell-checker or computer. A spell-checker cannot interpret text, and will only check spelling. Missing words are not spotted.

Gastel and Day (2016) mentions the first English edition of the Bible, the so-called King James version, printed in 1631. The seventh commandment appeared as: "Thou shalt commit adultery". There may be some speculation about the popularity of this edition, but one thing is certain: a computer spell-checker would never have spotted this error.

Second, do not do the proofreading by yourself. Ask someone to help. This is because you are too familiar with the text by this stage, and therefore unlikely to spot missing letters, mix-ups and similar errors. You may notice missing words or sections but otherwise your ability to spot errors will be low. Asking someone else will greatly reduce such errors being overlooked.

One of you should take the original (the manuscript), and read it aloud. The other person should carefully follow the proofs to see if it matches the original. Do this twice — it may seem superfluous, but remember: any mistake that slipped your attention will remain there forever. Do not underestimate the "annoyance potential" of such small errors; help yourself to ensure a less stressful life.

Take special care with numbers, tables, and figures, symbols, equations, unusual expressions, scientific names. Remember, the people working on your manuscript at this stage are not scientists; they are "keyboard operators". They will not be able to spot if a decimal point error slipped through, if a column content is transposed, a scientific name is misspelled, etc. You are the only one who can spot such mistakes. And do not trust the computer — surprising errors can occur.

On finding an error, the traditional method used to be to mark its position in the text, and indicate the correction in the margin. For this, standard proofreading marks were used by people in the publishing and printing industry worldwide. Electronic proofreading systems seem to be increasingly used, in which the author must use the comments and correction tools in Adobe Acrobat. In some cases, you will have to return the corrected pdf file; for other journals, you will have to login into the proofing system of the journal on the Internet, and make the corrections directly. If your Internet link is not good enough for this, you can always download the pdf file, and work on that, then send the corrected file back by e-mail.

You should also check the placing, sequence and quality of tables and figures. Are the figure sizes large enough? They are rarely too big — but frequently too small, because the technical editor, from a desire to accommodate more papers on the limited number of pages available to the journal, will try to reduce size a little too much. Also, check the orientation of the figures, their sharpness and readability. If you are dissatisfied with them, you can request replacement or a size change, but indicate the reason. In that case, request the second proofs, too. For text changes, this is not necessary. Reply to any queries put to you, if there are any. These will usually concern word choices, sometimes missing data, numbers, etc. If there are any, correct them carefully.

At this stage, making changes can be very costly, because the whole issue is now typeset. As a rule, there cannot be any changes. If there is a missing paragraph that was omitted by the printer, they will have to correct it — this is not your error and, thus, you will not be expected to meet the costs. However, if you omitted a paragraph from the final manuscript, and want to include that in the printed paper, this can be very expensive. This cost will routinely be charged to the authors.

Additionally, check and update the reference list, especially the references that were, at the time of submission, "in press". Add the final data if available; these usually comprise simply the year of publication, the volume and page numbers.

If not done earlier, you must now sign the copyright transfer form. If you have to pay any charges, this usually happens at this stage, too.

26

The Paper is Published. Now What?

Once you have published a paper, you can sit back and wait for recognition and world fame to arrive. It may be a long wait. Until that happens, there are a few things that are necessary or advisable to do.

First, remember to provide the necessary depository copies, or to deposit the electronic file in the appropriate digital repository. Sending copies of your output to your funding organisations is often not just a courtesy, but an obligation, and is specified in the research contract. Do not neglect this. If more physical copies are needed than you have, print and mail them. Foundations, especially private foundations, often collect the outputs that emerge from projects done with their support. This may be an important way of documenting the usefulness of their existence.

Of course, if you are the corresponding author, you should make sure that all your co-authors have copies of the published paper. Your institutional library may also have an archive containing the outputs from your unit or department. Give (or send) a copy to the librarian. It is polite (and, therefore, a good idea), to give copies to people who appear in the paper by name, either as sources of personal communication, or who you acknowledged for helping you along the way to publication. Likewise, it is a friendly gesture to send a copy to cited authors — but use this option with restraint: maybe to authors of primary papers you cited, but not to those of books or reviews.

You can also use the published paper as a networking tool, giving copies to your colleagues, group leaders, friends, parents, rivals, supervisors, etc. Think about other people or organisations who can benefit from the information you published. These might be non-governmental organisations, museums, collections, schools, field stations and the like. They share one thing in common: difficulties in obtaining such primary scientific literature.

Reprint Requests

The custom of asking for copies of papers is dying, or may be a thing of the past already. When papers were mostly printed, authors received a limited number of printed copies (reprints) and anyone could ask for a copy. Practically, you never had to buy a reprint published by someone else — reprints were part of the toolkit that scientists used to network with each other. To some degree, the electronic replacement of this practice still thrives. You will receive requests for copies from fellow scientists, institutions, students, etc. It rarely costs you much to fulfil these requests, and I suggest you do.

Archiving Your Paper on a Personal Website

Uploading an electronic copy of your paper, either on your personal website or that of your organisation, may increase the availability of your paper to the wider public. This option may be useful — but observe any legal limitations that may exist. Some publishers allow you to upload your paper — but only as an accepted, non-typeset manuscript version. Others are not so restrictive. You may also be asked to deposit your work in a publicly available repository. Such options are useful, because it increases the chance that others will find your paper. However, this depends on where the work was originally published. Check the relevant rules and regulations, including the copyright agreement.

27

The Process of Writing a Conference Proceedings Paper

A conference proceedings paper is a very common kind of publication. However, its usefulness requires scrutiny. Therefore, your first important question, on receiving an invitation to contribute to the proceedings of a conference you have attended, should not be "how to write a conference proceedings paper?" The real question is: should you write a conference proceedings paper at all?

Often, a conflict exists: a conference proceedings paper is not usually a primary publication. Therefore, you earn little publication benefit from publishing such a paper. Still, travel support is often linked to presenting material at a conference (and rightly so: if no one presented anything, everyone would be expecting others to fill the conference with content), and once much of the work is done, what harm comes from turning this into a conference proceedings paper?

Additionally, there may be pressure of a different type and magnitude to publish your contribution in the conference proceedings. This pressure may come from your university supervisor: she expects you to perform at "her" conference, and of course, to submit your contribution to the proceedings. Alternatively, your own university might be organising the conference, and what is more natural than to submit your contributions to the proceedings? Doing otherwise may seem a betrayal of your own workplace. Thus, a refusal is not always an option, even if, at heart, you would like to do so.

The problems with conference proceedings are multiple. The purpose, it seems, of a volume of conference proceedings is not really to inform the outside world about the results presented at the meeting. All too often, the proceedings volume is a kind of monument to the conference, evidence for the funders or organisers that the conference took place. So, frequently, its main purpose is not to publish new scientific discoveries.

Additionally, this monument is constructed on a shoestring: the leftover money from the conference. This is almost never enough to engage a professional editor or printer. Therefore, the editors (often the conference organisers) step in to provide this service. They may serve the real interest of the meeting perfectly in this respect — and the significance of a conference is, in the eyes of an outsider, positively correlated with the size of the proceedings volume. A big book equals a big, important conference. Of course, everyone claims this is not so — yet the logic somehow survives. Unfortunately, there are several undesirable consequences.

The first is "soft" peer review. As a general perception, the participants, by paying to attend at the conference, have somehow bought the right for their contribution, presented at the conference, to be included in the proceedings volume. As this is still part of the scientific literature, the editors try to provide comments on every manuscript. The aim of this review, however, is not to select the best contributions offered. The editors try to help the authors to improve their manuscript but rejection is rarely on the horizon. Exceptions exist, but acceptance is commonplace in the world of conferences; therefore, the peer review is rather "soft".

The second hindrance is the page limit on individual contributions. Funds are always tight, and everyone expects to be included. Thus, the total number of pages available has to be evenly divided — more or less. Consequently, a conference proceedings paper always has severe limits: the length of text, number of figures, tables, and references are typically restricted. The overall limit can vary between five and 15 pages. It is rarely more, which virtually precludes a full primary publication — there is never enough space to substantively present the material and methods, for example. Do not spoil your chances of such a publication by trying to include all your primary data in a volume of conference proceedings.

Thirdly, the usual lack of money precludes the engagement of a professional editorial service. Editing, typesetting, etc. are usually done by amateurs. You cannot rely on them in the way you can expect a professional editorial and printing service from a journal. Be *very* careful and conscientious, especially with the proofreading. You are your own quality controller.

The general structure of the conference proceedings paper follows that of the primary paper. Frequently, though, the summary/abstract is not included, because that was originally published in the conference book of abstracts. Otherwise, the structure follows that of the primary scientific paper: there is an introduction, material and methods, results, and a discussion. The material and methods must be short — there is no space to present all of the necessary details. Try to point to other publications that provide more detail, rather than describing them here. The results also need to be short, and to the point. There is not a lot of space for discussion, either. Try to concentrate on the results. Above all, remember that this is rarely a valid primary publication, and do not publish your full set of results. Always retain the (copy)right to publish it later as a valid, primary publication. For reasons mentioned above, this will not be very difficult — conference proceedings papers are usually short, and do not allow you to present all the necessary detail.

I should add that, in several fields (for example, engineering), regular and well-attended conferences publish the full, written material in regular proceedings *before* the conference starts. However, this mechanism usually has the characteristics of peer-reviewed journals: authors must submit the full, written version of their planned contributions beforehand, which are evaluated by experts and accepted or not. These can be treated as full primary papers.

28

Reviews and the Process of its Writing

A review is an evaluation of published knowledge. Hence, reviews are not primary publications, because they do not contain new, unpublished information. However, this does not make them of little value; on the contrary, a good review can be very influential. Reviews are also usually long and detailed — up to 50 printed pages. In general, reviews are the first ports of call for newcomers to a subject area, and are used broadly by many to keep themselves up to date with progress in their field.

For a review, there is no commonly accepted general structure. However, if you want to think about it in relation to the structure of the primary article, a review does not usually contain a Material and Methods section (but see later in this chapter), and it contains few new results. It does contain an extended Introduction and Discussion; conclusions and recommendations are also part of a good review.

All involved realise that writing a review is a major undertaking. Thus, the usual way of going about a review is different from the "normal" manuscript process. Authors are not requested to write and submit a complete review, risking rejection after several months' work. So, if you intend to write a review, it is expected that you contact the editor of an appropriate publication in advance, by writing a proposal suggesting that you produce the review you have in mind.

This proposal should be relatively short (3-4 pages maximum for a long review), and contain the following elements:

- Your arguments about the topic — why is this special area, or problem ripe for a review now? Reasons could include a lack of recent review, important new information, a shift in the main paradigms, that it is an emerging new field or sub-field, or some recent trend or event that makes the review topical and important. Beware — the fact that this topic has not been reviewed before is rarely a sufficient argument.

- Next, you should present your, or the team's credentials: why are you the best one(s) to write this review? At this point, you should be able to demonstrate some relevant experience in the primary field. A past record of being able to write well is also received favourably by editors.
- The editors would also like to have as much information as possible about the intended scope and structure of the suggested review. Indicate the extent and limits of the review. Is it going to be focusing on theoretical or applied aspects? Will it be covering or emphasising any specific habitat, geographical area, group of organisms, methods or a phenomenon? How do you intend to organise the information? How do you intend to collect material for the review? The more detail you can give at this stage, the better.

This letter of approach should not be longer than about 3-4 pages, but all the above must be covered. The editor(s) will assess this letter, and they will contact you with their decision: a rejection, an invitation, or they may request further clarifications. An invitation rarely takes the form of a promise to publish, as editors do not like to commit themselves to publication before they have had an opportunity to read the final manuscript. Reviews are also peer-reviewed, but the scope is slightly different. Such peer reviews are more of a set of suggestions and modifications, and not profound criticisms. Nevertheless, there is still no guarantee that your review will be accepted for publication.

Writing the Review

The usual restrictions apply with respect to manuscript submission. Approach only one editor at a time, and do not try to negotiate "the best possible deal" by playing one journal off against another one. If your approach is not successful at one journal, you can then try another forum.

The style of review will have to consider the expected readership — and this is always wider than that of primary papers, and includes peers, colleagues, and students. Thus, the style should be general and expansive, with the non-expert in mind. Explain major concepts in plain language. Use summary tables and figures, if appropriate — it often is.

As the review relies on already-published information, be aware of potential copyright issues. If you want to use a published figure or table, even if you combine several tables into one, you must obtain permission from the copyright holder, usually a publisher (unless the figure or table is openly licensed). This process will take some time — so think about it early in your writing and literature search. Try not to cite word by word — use your own expressions when summarising or presenting others' results. If you want to cite word-by-word, a few sentences, and a maximum of one paragraph can be reproduced using quotes and a reference to the original. Anything longer needs not only a proper citation of the source, but also permission to use it (unless it is openly licensed).

Read Papers, Not Abstracts

An abstract is already a re-interpretation of the most important results, even if by the authors themselves. Do not just use their interpretation — read the paper, and make the summary your own. Your review will suffer if you cite a paper that you have not seen — do not do it.

Citation maps are a useful tool during the initial orientation phase. Do not rely on this method only — use computer searches, check the Web of Science website, or other literature databases. You will notice that they never overlap completely.

To organise the structure of your review, do a mind map. This will help you to create a proportional structure — you can decide which sub-topics you want to give more emphasis, and allocate your reading efforts accordingly. It is pointless to spend too much time searching for information, reading and writing on a sub-topic of little importance.

Learn how to use Boolean operators. These allow you to link different words and create a structure for your search — a search string. Do not try, however, to arrive at your final set of relevant papers by constructing one long, perfect search string. You can do sequential searches, gradually narrowing in on your target. You can start with a wide and general search, and refine this by searching the hits generated by your first search using more specific terms. Alternatively, you can do several one-word or concept searches, and then combine them in various ways. It is always easier to do several smaller searches than one big one — the latter is rarely perfect. Stepping back one step is much wiser than having to start all over again.

Just as clarity and explaining difficult concepts are critical, conclusions are also of great importance in a review. The reader expects you not to merely list them but to give guidance, to assess, and to evaluate. This is your chance to give direction to your field. Where are the most important challenges? Which things are easy to do (but necessary) and what is not easy to approach now? Why not? Are there conceptual or methodological obstacles? Is the body of evidence still too small? Are there too many controversies?

You can follow a chronological order during your review — in fact it is almost essential to do so if you want to track the development of a theory or idea. This is not the only possible structure, however. You can write a "state-of-the-art" type of review, in which the detail, or sequence, is determined by the number of relevant papers. You might want to detail controversial areas, or follow another non-linear "storytelling" structure. It is up to you.

29

Book Chapters and the Process of Writing them

Books have become very important tools of information and learning. In the view of many, a scientific book is an authoritative source of information, written mostly for those who know little about the subject. Writing a book is a very large and complex task. Traditionally, writing scientific books has been the realm of the "real expert", usually near the peak of an illustrious and productive career.

Today, however, fewer and fewer people write single-authored scientific books. The reasons for this are manifold, but the two most important ones are the huge increase in the amount of information, and the increasingly limited time available to researchers to focus on such a task — or so most people claim.

In the absence of knowledgeable, willing and able single authors, there are more and more books that are written by a group of experts. In some cases, every chapter is written by a different set of authors. Such a book is then coordinated by one or more scientific editor(s). Receiving an invitation to contribute to a book is flattering, because it means that you are considered by your colleagues to have a certain comprehensive knowledge as an expert in your chosen field of research, and they trust you to impart this knowledge to others in an accessible way. However, do not let yourself be led by your vanity into such a venture. Once your pride subsides a little, consider a few things very carefully and critically before accepting the invitation.

The success of a book depends on content, form, and marketing.

Let us assume that you are confident that you can write an excellent chapter alone, or in a team, about the topic you were asked to write about. Of course, if you do not feel this way, do not accept the invitation.

However, if you do feel this way, this is far from sufficient to give a positive reply. Writing a chapter is still a considerable piece of work and, if the entire book is unsuccessful, your contribution will not get the attention it should. So: what are the necessary conditions for a successful, multi-authored book?

The first is organisation. The editors will have a large and complicated task: to make sure that many authors complete their given contribution on time, and to a reasonable standard. This is a huge organisational challenge. Additionally, there must be some stylistic editing done, as it is guaranteed that authors will have very different approaches to their topic, and different styles. An editor must be an efficient organiser, a diplomat, and a conscientious controller of a complicated venture. This role is very important, as it keeps the project progressing on time, and this requires a lot of time, effort, and scientific — as well as diplomatic — skills. Only join an author team if you know the editor, preferably personally, and are favourably impressed by her abilities as scientist, writer, and organiser.

If there are multiple authors, it also becomes a question of "company". One good chapter will not make the reputation of the book, nor assure its success, so you must be sure about the quality of your co-authors. Find out about the other authors who were asked to contribute, and only join the team if you trust them to deliver high-quality manuscripts.

Finally, a set of high-quality manuscripts delivered on time are not enough: books have to sell, and for this, they must be printed and distributed. Only a good, experienced publisher with a record of producing high-quality books would provide a reasonable chance of success. Only join the team if the suggested publisher is well known in the field, and has a reliable record of producing such books, as well as a global distribution chain.

If all three conditions are met, then consider the available time, your own engagements and, if you have the time, go ahead: good books are rare, and are in high demand. The task is to write for the novice. Book chapters are also non-primary publications, so never include previously unpublished information.

During the actual writing, use the experience gained when writing reviews, or refer to the points mentioned in the previous chapter. Assume even less of the reader's background knowledge than for a review.

Frequently, people reading books know next to nothing about the subject and the function of the book is to provide them with this knowledge. Write in plain language, use lots of examples, figures, pictures, and tables. You will almost certainly use the work of others — so dealing with copyright is almost inevitable. The publisher can help you with this, but do not leave it to the end.

30

Major Aspects of the Scientific Style

Never fear big words. Long words name little things. All big things have little names, such as life and death, peace and war, or day, dawn, night, love, home. Learn to use little words in a big way — it is hard to do. But they say what you mean. When you don't know what you mean, use big words: they often fool little people.

SSC Booknews, 1981, quoted in Day, 1989

The most important principle of the scientific style is simplicity and clarity. New information is not easy to understand. You also recall that complicated coding can stand in the way of decoding (or, to put it another way, understanding). During reading, the reader decodes and interprets the information; interpersonal differences, reading conditions, and the predisposition of the reader will lead to varying interpretations. The bigger these differences are, the higher is the chance of misinterpretation. Note that I am not talking about legitimate interpretational differences here (the writer has no monopoly on the "proper" interpretation of her writing), but misunderstandings.

Readers expect context on the left, new information on the right (Gastel and Day 2016), so organise your sentences to meet this expectation. The subject of a sentence should be followed by a verb as soon as practicable. The beginning of the sentence is the topic position, letting the reader know the topic of the sentence. The end is the stress position, and this is the place for information that requires emphasis. In one sentence, try to communicate one point.

The reader cannot understand new information without context — provided by earlier, known, published information — so begin with old information, and finish with new information, also from sentence to sentence. This way you provide context before asking the reader to consider new information.

Be careful with similes and metaphors. Use them rarely and carefully. Otherwise, unintentionally funny results can emerge, such as this classic: "A virgin forest is a place where the hands of man have never set foot" (Gastel and Day 2016).

Be mindful of sentence structure. Day (1989) mentions a questionnaire, sent to fire-brigade leaders through the UK. The head of a brigade in Hampshire read the question carefully.

Question: How many people do you employ, broken down by sex?

Answer: None. Our problem here is booze.

At appropriate places in this book, I mentioned the tenses to use when composing different sections of a paper. Mostly, you will only use two of them: the simple present and the simple past. There are two main types of statements in a scientific paper: already-known facts, and new discoveries. The former should be mentioned using the simple present tense (and a reference to its discoverer). The latter should appear in the simple past (and is supported by experimental evidence: numbers, tests, figures, tables, etc.). Grammar occasionally requires a different tense, but typically, the grammatical structure of a scientific paper is simple, rarely using other than the above two tenses. As a Kenyan course participant once put it to the class: "Do not use the complicated present".

Current writing style emphasises the importance of the use of the active voice; the passive should generally be avoided, especially in American English usage. The first person "I" is not wrong — although it is rarely used, because most published work is done by teams. Be mindful, though, of when to use it, and avoid the much-cited ridiculous statement by (as she was then) Mrs Thatcher: "We are a grandmother". Beware of self-cancelling words, fillers, and what Gastel and Day (2016) call "mumblespeak" (see a list of jargon words and their simple equivalents in Day and Gastel, 2016, Appendix 2).

Jargon fulfils an important function in communication. However, its usefulness is very context-dependent, and mostly limited — this is one of its very functions. The same holds for acronyms. These have become the darlings of complicated bureaucracies (for example, the European Union administration), who seem to revel in them. The use of acronyms also fulfils some of the functions of jargon. Those who are "in" will understand them, and those who are "out" are, rightly, baffled. Therefore, everyone pretends to understand them. Do not follow their example. If you use an abbreviation, provide a definition or write it out in full at first mention, followed by the abbreviation or acronym in parentheses. Subsequently, you can just use the acronym.

These are only a few pointers. More detailed advice on scientific style can be found in several books (e.g. Clymo, 2014; CBE, 1994; Turabian, 2007). As a final, humorous resource, Day and Gastel (2006) offer the enclosed list of the ten most common mistakes that it is claimed non-native writers make in their use of English. Note that the list commits the errors to be avoided (see Box 17).

Box 17. Day and Gastel's (2006) ten most common mistakes that it is claimed non-native writers make in their use of English

1. Each pronoun should agree with their antecedent.

Here the appropriate use of singular and plural is to be observed: the word "antecedent" is in singular, to the appropriate pronoun would be "its".

2. Just between you and I, case is important.

This is tricky because there are not so many cases in English, and not all of them are different in the different cases. For example, the most common one, "you" has the same form.

3. Don't use no double negatives.

Many languages, for example those in the Latin language family, use double negatives. "Non faccio nulla" — in Italian means "I do nothing", but translated word-by-word, this is "I do not do nothing". English does not use double negatives, so the correct form is either "I do not do anything", or better: "I do nothing".

4. A preposition is a poor word to end a sentence with.

Note, however, the oddity Winston Churchill famously points out: "This is a rule up with which I will not put."

Churchill refers to English verbs that, while superficially in the form of a verb with a preposition behind, have acquired a new meaning, and are now inseparable. "To put" is synonymous with "to place", or "to position", but "to put up" means "to tolerate". Here, the "up" is always immediately after the "put", and thus the "up" actually can stand at the end of a sentence and still be correct: "This outcome is one with which we cannot put up."

5. Verbs has to agree with their subject.

In long, complex sentences it is not always obvious what is the appropriate subject to a verb, so this is a frequent mistake — not only by non-native English speakers. Be careful.

6. Remember to never split an infinitive.

This is, actually, a much debated rule, with many famous writers protesting against it. However, in most cases, it would be an error to split the infinitive.

7. When dangling, don't use participles.

Words with an ending of -ing or -ed are called participles. They can be present (breaking, going, drinking) or past participles (broken, gone, drunk). The participle goes with the noun closest to it, either directly preceding or following it and the words which go with it in the sentence. The antecedent—that is, the noun to which the participle refers—must be clear to the readers in order for them to understand what's being said. Otherwise, an action may be subscribed to the wrong player. That's called a "dangling participle," because it's left "dangling" without a clear antecedent. Consider the example: "The robber ran from the policeman, still holding the money in his hands." It is likely that the robber was holding the money — but as the word "policeman" is closest to the participle, readers get the wrong meaning. Corectly, it should be: "The robber, still holding the money in his hands, ran from the policeman."

8. Join clauses good, like a conjunction should.

A conjunction cannot be used with just one clause. Conjunction joins TWO clauses, usually written as one sentence.

Mistake: That I didn't know what to do.

Correct: I explained that I didn't know what to do.

9. Don't write a run-on sentence it is difficult when you got to punctuate it so it makes sense when the readers read what you wrote.

10. About sentence fragments.

There is no need to comment on points 9 and 10 — they are self-explanatory.

A Final Note

As I wrote at the beginning of this book, this is about the *form, not the substance* of science. Be aware that *nothing* will help you, neither this book nor any other, if you do not have substance. Also, do not go into science for fame, money, or power. Most former, or currently active, scientists are not famous, rich or influential either. Be a scientist because of the joy and satisfaction of doing science — there is no other worthwhile reward. Success, money, or fame will either elude you, or will prove ephemeral and hollow. By the time you discover that this is indeed so, it will be too late: you will have spent your life doing something that you did not really like.

I bid you farewell with the words by Ferenc Karinthy, a Hungarian writer from the early 20th century, famous in his country for a witty book, *This Is How You Write* (Karinthy, 1912) satirising his contemporary colleagues:

> The new recruits are practicing target shooting, under the instruction of a corporal. It is not going very well. The angry corporal is cursing the recruits, then snatches the rifle from the hands of one, after he had missed the target again.
>
> - Oh, you fools—shouts he—only if you could shoot straight! Give me that rifle! Look!
> He aims with confidence, and pulls the trigger. The shot misses the target. He is dumbfounded for a minute. Then he morosely turns to one of the recruits:
> - This is how *you* shoot!
> He aims again. Misses. Turns to another one:
> - And this is how *you* shoot!
> On it goes. Finally, the ninth shot hits the target. He proudly puffs out his chest:
> - And this is how *I* shoot!

> The ninth shot is still to come. The hands of the corporal are shaking, but he can see the target just a little more clearly.

Remember — it is often easy, and sometimes educational, to see and point out others' errors. Sooner or later, you also must start "shooting". I wish you strength, perseverance, and good luck in your career as a publishing scientist.

Literature Cited

Audusseau H, Baudrin G, Shaw MR, Keehnen NLP, Schmucki R, Dupont L, 2020. Ecology and genetic structure of the parasitoid *Phobocampe confusa* (Hymenoptera: Ichneumonidae) in relation to its hosts, *Aglais* species (Lepidoptera: Nymphalidae). *Insects*, 11, 478, doi: 10.3390/insects11080478.

Barrass R, *Scientists must write. A guide to better writing for scientists, engineers and students*. 2nd ed. Routledge, London & New York.

Cargill M, O'Connor P. 2013. *Writing scientific research articles*. 2nd. ed. Wiley-Blackwell, Chichester, UK.

CBE (Council of Biology Editors) 1994. *Scientific style and format. The CBE Manual for authors, editors and publishers*. 6th edition. Cambridge University Press, Cambridge, UK.

Cho H, Lee WY. 2020. Interspecific comparison of the fecal microbiota structure in three Arctic migratory bird species. *Ecology & Evolution*, 10, 5582–5594, doi: 10.1002/ece3.6299.

Cleveland WS. 1993. *Visualizing data*. Hobart Press, Summit, N.J., USA.

Cleveland WS. 1994. *The elements of graphing data*. CRC Press, Boca Raton, USA.

Clymo RS. 2014. *Reporting research. A biologist's guide to articles, talks, and posters*. Cambridge University Press, Cambridge, UK.

Davies RG, Eggleton P, Dibog L, Lawton JH, Bignell DE, Brauman A, Hartmann C, Nunes L, Holt J, Rouland C. 1999. Successional response of a tropical forest termite assemblage to experimental habitat perturbation. *Journal of Applied Ecology*, 6, 946-962, doi: 10.1046/j.1365-2664.1999.00450.x.

Day RA, Gastel B. 2006. *How to write and publish and scientific paper*. 6th edition. Cambridge University Press, Cambridge, UK.

De Grave S, Smith KG, Adeler NA, Allen DJ, Alvarez F, Anker A, Cai Y, Carrizo SF, Klotz W, Mantelatto FL, Page TJ, Shy J-Y, Villalobos JL, Wowor D. 2015. Dead shrimp blues: a global assessment of extinction risk in freshwater shrimps (Crustacea: Decapoda: Caridea). *PLoS ONE*, 10(3), e0120198.

De Maagd RA, Bosch D, Stiekema W. 1999. *Bacillus thuringiensis* toxin-mediated insect resistance in plants. *Trends in Plant Science*, 4, 9-13, doi: 10.1016/s1360-1385(98)01356-9.

Elek Z, Lövei GL, Bátki M. 2017. Sex-specific interaction of body condition and asymmetry in carabids in distinct urbanisation stages. *Community Ecology*, 8, 253-259. doi: 10.1556/168.2017.18.3.4.

Fox J, Petchey O. 2010. Pubcreds: fixing the peer review process by "privatizing" the reviewer commons. *Bulletin of the Ecological Society of America*, 1, 325–333.

Fricke EC, Tewksbury JJ, Rogers HS. 2019. Linking intra-specific trait variation and plant function: seed size mediates performance tradeoffs within species. *Oikos*, 128, 1716-1725, doi: 10.1111/oik.06494.

Gastel B, Day RA. 2016. *How to write and publish and scientific paper*. 8th edition. Greenwood, Santa Barbara, CA, USA.

Gausman J, Austin SB, Subramanian SV, Langer A. 2020. Adversity, social capital, and mental distress among mothers of small children: A cross-sectional study in three low and middle-income countries. *PLOS PLE*, 15(1), doi: 10.1371/journal.pone.0228435.

Godínez-Alvarez H, Ríos-Casanova L, Peco B. 2020. Are large frugivorous birds better seed dispersers than medium- and small-sized ones? Effect of body mass on seed dispersal effectiveness. *Ecology & Evolution*, 10, 6136–6143, doi: 10.1002/ece3.6285.

Greenop A, Cook SM, Wilby A, Pywell RF, Woodcock BA, 2020. Invertebrate community structure predicts natural pest control resilience to insecticide exposure. *Journal of Applied Ecology*, 57, 2441-2453, doi: 10.1111/1365-2664.13752.

Harzing A-W. *2010. The publish or perish book: Your guide to effective and responsible citation analysis*. Tarma Software Research Pty., Melbourne, Australia.

Himanen SJ, Nerg A-M, Poppy GM, Stewart CN, Holopainen JK. 2010. Abiotic stress and transgenics: implications for reproductive success and crop-to-wild gene flow in Brassicas. *Basic and Applied Ecology*, 11, 513-521.

Hirsch JE. 2005. An index to quantify an individual's scientific research output. *Proc. Natl Acad Sci U S A.*, 102, 16569-16572.

Hunt R. 2001. Trying an authorship index. *Nature*, 52, 187.

Imboma TS, Gao D-P, You M-S, You S-J, Lövei GL. 2020. Predation pressure in tea (*Camellia sinensis*) plantations in southeastern China measured by the sentinel prey method, *Insects*,11, 212, doi: 10.3390/insects11040212.

Karinthy F. 1912. *Igy irtok ti* [This is how you write]. Athenaum Publisher, Budapest, Hungary.

Kelly S. 2018. The continuing evolution of publishing in the biological sciences. *Biology Open*, bio037325, doi: 10.1242/bio.037325.

Klimek-Kopyra A, Dłużniewska J, Ślizowska A, Dobrowolski JW. 2020. Impact of Coherent Laser Irradiation on Germination and Mycoflora of Soybean Seeds—Innovative and Prospective Seed Quality Management. *Agriculture*, 10, 314, doi: 10.3390/agriculture10080314.

Laughlin RB, Pine D. 2000. The theory of everything. *Proceedings of the National Academy of Sciences of the USA*, 97, 28-31.

Magura T, Lövei GL, Tóthmérész B. 2010. Does urbanisation decrease diversity in ground beetle (Carabidae) assemblages? *Global Ecology and Biogeography*, 19, 16-26, doi: 10.1111/j.1466-8238.2009.00499.x.

Magura T, Lövei GL. 2019. Environmental filtering is the main assembly rule of ground beetles in the forest and its edge but not in the adjacent grassland. *Insect Science*, 26, 154-163, doi: 10.1111/1744-7917.12504.

Masaki T, Nakashizuka T, Niiyama K, Tanaka H, Iida S, Bullock JM, Naoe S. 2019. Impact of the spatial uncertainty of seed dispersal on tree colonization dynamics in a temperate forest. *Oikos*, 128, 1816-1828, doi: 10.1111/oik.06236.

McMahon CR, Burton HR, Bester MN. 2003. A demographic comparison of two southern elephant seal populations. *Journal of Animal Ecology*, 2, 61–74, https://besjournals.onlinelibrary.wiley.com/doi/pdfdirect/10.1046/j.1365-2656.2003.00685.x.

Mingers J, Leyesdorf L (in press). A Review of Theory and Practice in Scientometrics. *European Journal of Operational Research* — Available at: https://arxiv.org/ftp/arxiv/papers/1501/1501.05462.pdf.

Murray BR, Fonseca CR, Westoby M. 1998. The macroecology of Australian frogs. *Journal of Animal Ecology*, 7, 567-579, doi: 10.1046/j.1365-2656.1998.00217.x.

Oedekoven MA, Joern A. 2000. Plant quality and spider predation affects grasshoppers (Acrididae): food-quality-dependent compensatory mortality. *Ecology*, 81, 66–77.

Oguz F, Koehler W. 2016. URL Decay at Year 20: A Research Note. *Journal of the Association for Information Science and Technology*, 7, 477–479.

Pechan T, Cohen A, Williams WP, Luthe DS. 2002. Insect feeding mobilizes a unique plant defense protease that disrupts the peritrophic matrix of caterpillars. *Proceedings of the National Academy of Sciences*, 99, 13319-13323.

Ratnoff OD. 1981. How to read a paper. *In: Warren KS, Ed. Coping with the biomedical literature*. Praeger, New York, NY, USA, pp. 95-101.

Reckinger C, Colling G, Matthies D. 2010. Restoring populations of the endangered plant *Scorzonera humilis*: influence of site conditions, seed source, and plant stage. *Restoration Ecology*, 18, 904-913, doi: 10.1111/j.1526-100X.2009.00522.x.

Resnik DB. 2011. A troubled tradition: it's time to rebuild trust among authors, editors and peer reviewers. *American Scientist*, 99, 24–27.

Simkin MV, Roychowdhury VP. 2003. Read before you cite! *Complex Systems*, 14, 269–274.

Skou JC. 1957. The influence of some cations on an adenosine triphosphatase from peripheral nerves. *Biochimica Biophysica Acta*,1000, 439–46.

Smith R. 2006. Peer review: a flawed process at the heart of science and journals. *Journal of the Royal Society of Medicine*, 99, 178–182, doi:10.1258/jrsm.99.4.178.

Sopinka NM, Coristine LE, DeRosa MC, Rochman CM, Owens BL, Cooke SJ. 2020. Envisioning the scientific paper of the future. *FACETS*, 5, 1–16, doi: 10.1139/facets-2019-0012.

Tscharntke T, Hochberg ME, Rand TA, Resh VH, Krauss J. 2007. Author sequence and credit for contributions in multiauthored publications. *PLoS Biology*, 5(1), e18, doi: 10.1371/journal.pbio.0050018.

Tufte ER. 1990. *Envisioning information*. Graphics Press, Cheshire, Connecticut, USA.

Tufte ER. 1997. *Visual explanations*. Graphics Press, Cheshire, Connecticut, USA.

Tufte ER. 2003. *The visual display of quantitative information*. 2nd ed. Graphics Press, Cheshire, Connecticut, USA.

Tufte ER. 2006. *Beautiful evidence*. Graphics Press, Cheshire, Connecticut, USA.

Turabian KL. 2007. *A manual for writers of research papers, theses, and dissertations*. 7th edition, University of Chicago Press, Chicago, IL, USA.

Uchmanski J. 2019. Retraction: Algorithmicity in biology. *Methods in Ecology & Evolution*, 10, 1628.

Valiela I. 2009. *Doing science*. 2nd ed. Oxford University Press, Oxford, UK.

Weber EJ, Katz PP, Waeckerle JF, Callaham ML. 2002. Author perception of peer review: impact of review quality and acceptance on satisfaction. *JAMA*, 287, 2790–2793.

Whitfield DP. 2003. Predation by Eurasian sparrowhawks produces density-dependent mortality of wintering redshanks. *Journal of Animal Ecology*, 72, 27–35, doi: 10.1046/j.1365-2656.2003.00672.x.

Permissions

All chapters in this book were first published in WPSP, by Open Book Publishers by Gábor L. Lövei; hereby published with permission under the Creative Commons Attribution License or equivalent. Every chapter published in this book has been scrutinized by our experts. Their significance has been extensively debated. The topics covered herein carry significant information for a comprehensive understanding. They may even be implemented as practical applications or may be referred to as a beginning point for further studies.

We would like to thank the editorial team for lending their expertise to make the book truly unique. They have played a crucial role in the development of this book. Without their invaluable contributions this book wouldn't have been possible. They have made vital efforts to compile up to date information on the varied aspects of this subject to make this book a valuable addition to the collection of many professionals and students.

This book was conceptualized with the vision of imparting up-to-date and integrated information in this field. To ensure the same, a matchless editorial board was set up. Every individual on the board went through rigorous rounds of assessment to prove their worth. After which they invested a large part of their time researching and compiling the most relevant data for our readers.

The editorial board has been involved in producing this book since its inception. They have spent rigorous hours researching and exploring the diverse topics which have resulted in the successful publishing of this book. They have passed on their knowledge of decades through this book. To expedite this challenging task, the publisher supported the team at every step. A small team of assistant editors was also appointed to further simplify the editing procedure and attain best results for the readers.

Apart from the editorial board, the designing team has also invested a significant amount of their time in understanding the subject and creating the most relevant covers. They scrutinized every image to scout for the most suitable representation of the subject and create an appropriate cover for the book.

The publishing team has been an ardent support to the editorial, designing and production team. Their endless efforts to recruit the best for this project, has resulted in the accomplishment of this book. They are a veteran in the field of academics and their pool of knowledge is as vast as their experience in printing. Their expertise and guidance has proved useful at every step. Their uncompromising quality standards have made this book an exceptional effort. Their encouragement from time to time has been an inspiration for everyone.

The publisher and the editorial board hope that this book will prove to be a valuable piece of knowledge for students, practitioners and scholars across the globe.

Index

www.ingramcontent.com/pod-product-compliance
Lightning Source LLC
LaVergne TN
LVHW011513200926
848484LV00001B/33
* 9 7 9 8 8 8 8 3 6 0 3 9 2 *